Springer Series in
OPTICAL SCIENCES   107

Founded by H.K.V. Lotsch

Springer Series in
OPTICAL SCIENCES

The Springer Series in Optical Sciences, under the leadership of Editor-in-Chief *William T. Rhodes*, Georgia Institute of Technology, USA, provides an expanding selection of research monographs in all major areas of optics: lasers and quantum optics, ultrafast phenomena, optical spectroscopy techniques, optoelectronics, quantum information, information optics, applied laser technology, industrial applications, and other topics of contemporary interest.
With this broad coverage of topics, the series is of use to all research scientists and engineers who need up-to-date reference books.
The editors encourage prospective authors to correspond with them in advance of submitting a manuscript. Submission of manuscripts should be made to the Editor-in-Chief or one of the Editors.

Jesse Zheng

Optical Frequency-Modulated Continuous-Wave (FMCW) Interferometry

With 219 Illustrations

 Springer

Jesse Zheng
1980 E. 51 Ave.
Vancouver V5P 1V9
British Columbia, Canada

Library of Congress Cataloging-in-Publication Data
Zheng, Jesse.
   Optical frequency-modulated continuous-wave (FMCW) interferometry / Jesse Zheng.
      p.   cm. — (Springer series in optical sciences ; v. 107)
   Includes bibliographical references and index.
   1. Interference (Light)   2. Interferometry.   3. Light, Wave theory of.   I. Title.
II. Series.
QC411.Z44   2005
535.47´0287—dc22   2004056611

ISBN 0-387-23009-2   Printed on acid-free paper.

Printed in the United States of America.   (EB)

9 8 7 6 5 4 3 2 1   SPIN 11018957

springeronline.com

◆

To my loving wife,
Nancy

◆

Preface

Optical interference plays a prominent role in scientific discovery and modern technology. Historically, optical interference was instrumental in establishing the wave nature of light. Nowadays, optical interference continues to be of great importance in areas such as spectroscopy and metrology. Thus far, the physical optics literature has discussed the interference of optical waves with the same single frequency (i.e., homodyne interference) and the interference of optical waves with two different frequencies (i.e., heterodyne interference), but it hardly ever deals with the interference of optical waves whose frequencies are continuously modulated (i.e., frequency-modulated continuous-wave interference).

Frequency-modulated continuous-wave (FMCW) interference, which was originally investigated in radar in the 1950s, has been recently introduced in optics. The study of optical FMCW interference not only updates our knowledge about the nature of light but also creates a new advanced technology for precision measurements.

This book introduces the principles, applications, and signal processing of optical FMCW interference. The layout of this book is straightforward. Chapter 1 gives a short introduction to optical FMCW interferometry by considering the historical development, general concepts, and major advantages provided by this new technology. Chapter 2 focuses on the principles of optical FMCW interference. Three different versions of optical FMCW interference—sawtooth-wave optical FMCW interference, triangular-wave optical FMCW interference, and sinusoidal-wave optical FMCW interference—are discussed in detail. Moreover, multiple-beam optical FMCW interference and multiple-wavelength optical FMCW interference are also discussed by this chapter.

Chapter 3 introduces the optical sources for optical FMCW interference. Since, in practice, only lasers can be frequency-modulated optical sources, and the most frequently used ones are semiconductor lasers, this chapter first introduces the general principles of lasers, including the concepts of stimulated emission, population inversion, optical resonators, laser modes, and frequency modulation, and then discusses the characteristic features of semiconductor lasers. Chapter 4 introduces the optical detectors for optical FMCW interference. The operating principles of the commonly used semiconductor photodiodes, including PN photodiodes, PIN photodiodes, and avalanche photodiodes, and related issues, such as photodiode biasing, photocurrent amplification, and noise sources, are discussed in detail.

Chapter 5 discusses the coherence theory of optical FMCW interference, including the effects of the frequency bandwidth of optical sources, coherence of optical FMCW waves, and influence of the phase noise of optical sources. Chapter 6 first indicates the fundamental requirements and techniques for constructing optical FMCW interferometers and then gives some examples of optical FMCW interferometers, which are modified from the classical homodyne interferometers such as the Michelson interferometer, the Mach-Zehnder interferometer, and the Fabry-Perot interferometer.

Chapters 7, 8, and 9 deal with fiber-optic FMCW interferometers and fiber-optic FMCW interferometric sensors. Optical fibers are cylindrical dielectric optical waveguides and have been widely used in the fields of image transmission and optical communication, providing a low-noise, low-attenuation, low-cost, long-distance, and flexible light-propagation medium. The application of optical fibers and fiber-optic components to optical interferometers can make the interferometers compact, reliable, flexible, and more accurate. Moreover, with optical-fiber technology, more advanced detection techniques, more sophisticated interferometers, even "solid" interferometers (i.e., all-fiber interferometers) can be developed.

Chapter 7 first briefly introduces optical fibers and fiber-optic components and then presents some typical fiber-optic FMCW interferometers, including fiber-optic Michelson FMCW interferometers, fiber-optic Mach-Zehnder FMCW interferometers, and fiber-optic Fabry-Perot FMCW interferometers. Chapter 8 discusses the multiplexing technologies of fiber-optic FMCW interferometers. Four important multiplexing methods (frequency-division, time-division, time-frequency-division and coherence-division) and the related multiplexed fiber-optic FMCW interferometers are discussed in detail. Chapter 9 presents a number of advanced fiber-optic sensors based on optical FMCW interference, including fiber-optic FMCW interferometric displacement sensors, fiber-optic FMCW interferometric strain sensors, fiber-optic FMCW interferometric stress sensors, fiber-optic FMCW interferometric temperature sensors, and fiber-optic FMCW interferometric rotation sensors (i.e., fiber-optic FMCW gyroscopes).

Chapter 10 discusses the signal processing of optical FMCW interference. Three different methods for both frequency measurement and phase measurement are discussed in detail.

This book is based on the author's twenty-year research experience in this field. The author has tried his best to make this book clear, concise, and correct. All the contributors to optical FMCW interferometry involved in this book are given credit; however, inventors in other related fields are not always acknowledged.

This book is intended for scientists, engineers, and researchers in both academia and industry. It is especially suited to professionals who are working in the field of measurement instruments. It can also be used as a textbook of modern optical interferometry for senior undergraduate or graduate students.

I would like to express special gratitude to Dr. Hans Koelsch of Springer-Verlag New York for his great help in creating this project. Thanks also to my son, Jack, for his valuable help in the preparation of this manuscript. Finally, I would like to dedicate this book to my wife, Nancy, for her constant encouragement and support.

Vancouver, 2004   Jesse Zheng

Contents

.

Glossary

English Letters

A	gain of an amplifier
A_{21}	Einstein spontaneous emission coefficient ($A_{21} = 8\pi h\nu_{21}^3 B_{21}/c^3$)
B	magnetic field
B_{12}	Einstein stimulated absorption coefficient
B_{21}	Einstein stimulated emission coefficient ($B_{21} = B_{12}$)
c	speed of light in free space
C	capacitance of a capacitor
C_B	Brewster constant
D	detectivity of an optical detector
D	directivity of a fiber-optic directional coupler
D^*	specific detectivity of an optical detector
e	magnitude of an electronic charge ($e = 1.6 \times 10^{-19}$ C)
E	electric field
E	energy level
E_0	amplitude of an electric field
E_F	Fermi energy (or Fermi level)
E_g	energy band gap
f	wave number (or spatial frequency) of an optical wave
f_c	chopping frequency of an optical detector
f_e	occupational probability of the allowed states by electrons
f_h	occupational probability of the allowed states by holes
F	perpendicularly applied force
g	gain coefficient of an active medium
g_{th}	threshold gain coefficient
h	Planck constant ($h = 6.63 \times 10^{-34}$ J.s)
I	intensity of an optical wave
I	photocurrent from an optical detector
I_0	intensity amplitude of an optical wave
J	electric current density

k	propagation number of an optical wave
k	scale factor of a fiber-optic gyroscope
$\boldsymbol{k}$	propagation vector of an optical wave
k_B	Boltzmann constant ($k_B = 1.38 \times 10^{-23}$ JK^{-1})
l	length or distance
l_c	coherence length of an optical source
L_e	excess loss of a fiber-optic directional coupler
L_i	insertion loss of a fiber-optic directional coupler
M	multiplication factor of an avalanche photodiode
n	refractive index of a transparent material
n_e	effective refractive index of a single-mode optical fiber
n_{ex}, n_{ey}	effective refractive indexes of the HE$_{11}$^{x} mode and the HE$_{11}$^{y} mode
n_{ce}, n_{ve}	number of electrons in the conduction band and the valence band
n_{vh}, n_{ch}	number of holes in the conduction band and the valence band
N	population density in an energy level
NA	numerical aperture of an optical fiber
NEP	noise-equivalent power
OPD	optical path difference
OPL	optical path length
P	light power
P	pitch of a gradient-index rod
Q	Q factor of a laser cavity
$\boldsymbol{r}$	position vector
r_{63}	electro-optic coefficient of a crystal
R	coupling ratio of a fiber-optic directional coupler
R	reflectivity of a reflecting surface
R	resistance of a resistor
$\mathfrak{R}$	responsivity of an optical detector
s	speed of an object
t_c	coherence time of an optical source
T	temporal period of an optical wave
T_b	period of a beat signal
T_m	period of a modulation signal
V	electric voltage
V	normalized frequency of an optical fiber
V	Verdet constant
V	visibility (or contrast) of interference fringes or a beat signal

Greek Letters

α	angular frequency modulation rate of an optical wave
α	average loss coefficient of an optical resonator
β	gain factor of a semiconductor laser
β	open-loop gain of an operational amplifier
β	propagation number of a guided mode in an optical fiber
γ	complex degree of temporal coherence
Γ	self-coherence function
Δ	relative refractive index difference of an optical fiber
ε	electrical permittivity
ζ	mode coupling coefficient of a birefringent fiber
ζ	ratio of the electron-hole pair contribution to the photocurrent
η	coupling-loss coefficient of a birefringent fiber
η	phase modulation efficiency of a PZT tube fiber-optic phase modulator
η	quantum efficiency of an optical detector
θ_B	Bragg angle
θ_c	critical acceptance angle of an optical fiber
λ	optical wavelength in a medium
λ_0	central optical wavelength in free space
λ_0	optical wavelength in free space
λ_c	cutoff wavelength of a photodiode
λ_s	wavelength of a synthetic wave
Λ	beat length of a birefringent fiber
Λ	wavelength of an acoustic wave
ν	frequency of an optical wave
ν_b	frequency of a beat signal
ν_D	Doppler frequency shift
ν_m	frequency of a modulation signal
ν_s	frequency of a synthetic wave
$\delta\nu$	frequency bandwidth of an optical source
$\Delta\nu$	frequency modulation excursion of an optical source
ρ	radiation energy density
σ	standard deviation of a noise
τ	delay time of a signal beam with respect to a reference beam
τ	propagation time of an optical wave
τ	time constant of an optical detector
υ	speed of light in a medium

ϕ	phase of an optical wave
ϕ_0	initial phase of an optical wave
ϕ_{b_0}	initial phase of a beat signal
$\delta\phi_0$	phase noise of an optical source
ω	angular frequency of an optical wave
ω	cross-sectional radius of a laser beam
ω_0	central optical angular frequency
ω_b	angular frequency of a beat signal
ω_D	Doppler angular frequency shift
ω_m	angular frequency of a modulation signal
$\Delta\omega$	angular frequency modulation excursion of an optical wave

Chapter 1

Introduction to Optical Frequency-Modulated Continuous-Wave Interference

Optical interference results from the superposition of two or more optical waves. The color pattern on a film of oil on the surface of water is a simple example of optical interference. Optical interference has played a prominent role in scientific discovery and modern technology. In the early part of the nineteenth century, British scientist Thomas Young (1773–1829) did his famous double-slit experiment to establish the wave nature of light. Half a decade later, American physicist Albert A. Michelson (1852–1931) and chemistry professor Edward W. Morley (1838–1923) used the now well-known Michelson interferometer to prove the nonexistence of ether, clarifying a century of misunderstanding and leading to a new era of quantum physics.

Nowadays, optical interference continues to be of great importance in the areas such as spectroscopy and metrology. Using a wavelength as a scale, many physical quantities, such as optical wavelength, optical refractive index, distance, displacement, velocity, temperature, pressure, electric field, magnetic field, vibration amplitude, rotation speed, and, possibly, gravitational wave can be measured.

In physical optics, we have systematically investigated the interference of optical waves with the same single frequency (i.e., homodyne interference) and the interference of optical waves with two different frequencies, such as the optical waves from a Zeeman laser or those modulated by one or two acousto-optic modulators at different frequencies (i.e., heterodyne interference). However, we never discuss the interference of optical waves whose frequencies are continuously modulated (i.e., frequency-modulated continuous-wave interference) because of the lack of suitable frequency-modulated continuous-wave optical sources. This situation remained until the early 1980s, when the single-mode semiconductor lasers become available. The study of optical frequency-modulated continuous-wave interference not only increases our knowledge about the nature of light but also creates a new advanced technology for precision measurements.

In the following two sections, we will first briefly discuss the development history of optical frequency-modulated continuous-wave interferometry and then summarize the characteristic features of optical frequency-modulated continuous-wave interference. The aim of this chapter is to give readers an overall impression about this new technology before they read further.

1.1 Historical Development

Frequency-modulated continuous-wave (FMCM) interferometry was originally investigated in radar during the 1950s, where it was called frequency-modulated continuous-wave radar. *Radar* is an acronym for radio detection and ranging. It is used for locating a distant object by means of a reflected radio wave, as shown in Figure 1.1.

The fundamental principle of FMCW radar can be explained with the graphs in Figure 1.2. A narrow-bandwidth radio wave generated by a generator is modulated with a proper waveform (for example, a triangle wave). A small portion of the radio wave is introduced to a mixer as a reference wave; the majority, however, is transmitted toward the target by a transmitting antenna. The reflected signal wave (or echo) is collected by a receiving antenna (maybe the same one as the transmitting antenna) and coherently mixes with the reference

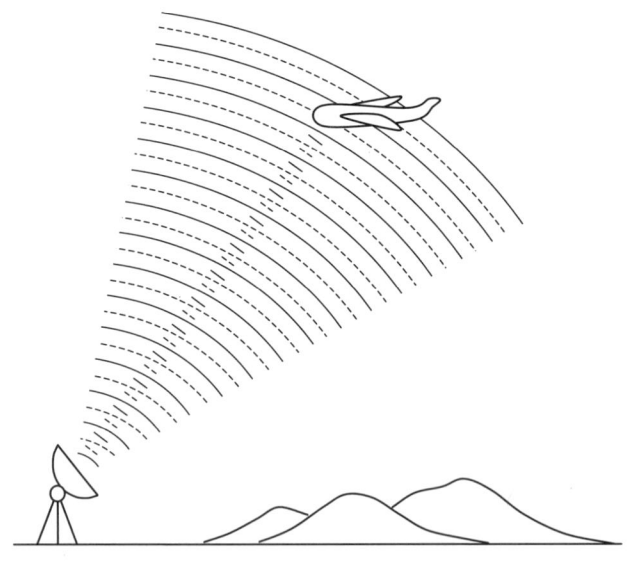

Figure 1.1. Application of radar.

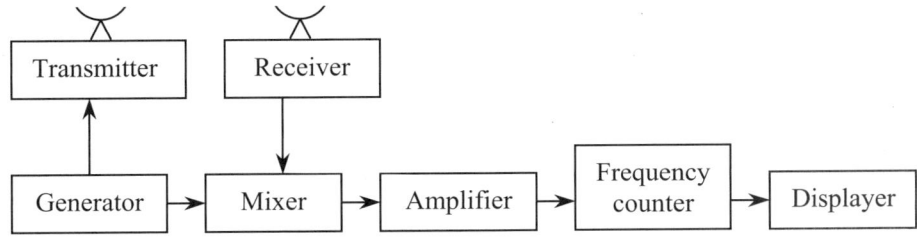

(a) Principle of FMCW radar.

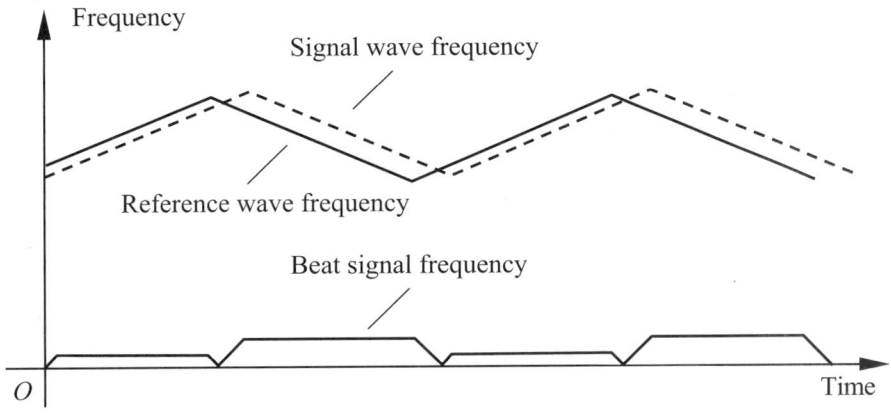

(b) Frequency-time relationships in FMCW radar.

Figure 1.2. Frequency-modulated continuous-wave (FMCW) radar.

wave at a mixer to produce a temporal signal (or beat signal) whose frequency is related to the propagation time of the signal wave and the frequency-modulation waveform. By measuring the frequency of the beat signal, the distance of the target can be determined.

The radar system with an optical wave usually is referred to as the light range finder (or *lidar*). The problem in FMCW lidar is that the coherence length of the available FMCW optical source is too short. (For instance, the coherence length of the commercial single-mode semiconductor laser is about 10 meters.) However, such a coherence length under most circumstances is long enough for an optical interferometer. Considering other superior features of FMCW interference, optical FMCW interference has received much interest in metrology.

The first paper regarding optical FMCW interference was reported by Brian Culshaw and Ian P. Giles in 1982 [15]. Since then, an extensive investigation on optical FMCW interference has been performed. Particularly, with the development of optical fibers and fiber-optic components, fiber-optic FMCW interferometers and fiber-optic FMCW interferometric sensors have been rapidly developed. To date, a large number of optical FMCW interferometric systems have been proposed and demonstrated. Some of them have already been in practice.

1.2 Optical FMCW Interference and Its Characteristics

Figure 1.3 schematically shows a simple optical FMCW interferometer. The configuration of the interferometer is exactly the same as in the classical parallel-beam Michelson interferometer except that the light source is a frequency-modulated semiconductor laser. A frequency-modulated laser beam is first collimated by a collimating lens and is then divided into two beams by a beam splitter. The reference beam propagates along the path l_1 and is reflected by a fixed mirror, while the signal beam propagates along the path l_2 and is reflected by a moving mirror. These two reflected beams are recombined by the same beam splitter to produce a beat signal, and the beat signal finally is detected by a photodiode. By measuring the frequency and the phase of the beat signal, we can determine the distance, displacement, and speed of the moving mirror.

Optical FMCW interference has the following major characteristic features, most of which are the advantages over the classical optical homodyne or heterodyne interference.

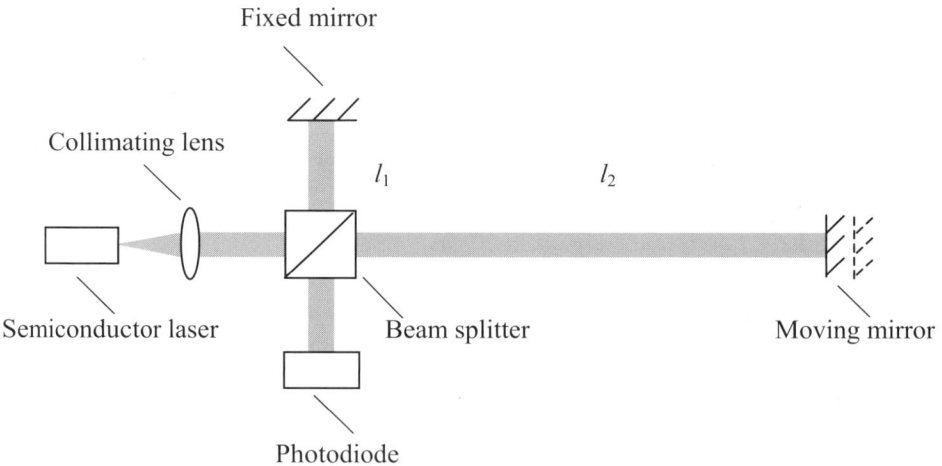

Figure 1.3. A simple optical FMCW interferometer.

(1) Optical FMCW interferometers usually measure a stable or quasi-stable target so that both the frequency and the phase of the beat signal can be used to acquire the information of the target. The frequency of the beat signal is related to the absolute distance of the target, while the variation of the initial phase (i.e., the phase shift) of the beat signal is related to the relative displacement of the target.

(2) The capability to measure the absolute distance is particularly important because the information can be recovered after an intentional or unintentional power interruption.

(3) The phase measurement generally gives a resolution thousands of times higher than the frequency measurement. Particularly, since the signal of optical FMCW interference is a dynamic signal (i.e., a continuous function of time), to calibrate the fractional phase, distinguish the phase-shift direction and count the number of full periods is quite easy. Therefore, compared with traditional optical interference, optical FMCW interference can offer a higher accuracy and a much longer measurement range.

(4) Optical FMCW interferometry is more varied than FMCW radar. Not just simply sending and receiving but also optical FMCW interference can be realized with various functional optical FMCW interferometers, such as the Michelson FMCW interferometer, the Mach-Zehnder FMCW interferometer, and the Fabry-Perot FMCW interferometer.

(5) Optical FMCW interference is well-suited for constructing fiber-optic interferometers and fiber-optic interferometric sensors. Optical fibers are cylindrical dielectric optical waveguides and have been widely used in the fields of image transmission and optical communication, providing a low- noise, low-attenuation, low-cost, long-distance, and flexible light-propagation medium. Fiber-optic components are miniature optical components that can perform various functions (such as splitting, combining, isolating, polarizing, etc.), the processes of which are performed inside the optical fibers. Application of optical fibers and fiber-optic components to optical interferometers can make the interferometers compact, reliable, flexible, and more accurate. In addition, with the fiber-optic technology, more advanced detection techniques, more sophisticated interferometers, and even "solid" interferometers (i.e., all-fiber interferometers) can be developed. Figure 1.4 shows the major components of fiber-optic FMCW interferometers.

(6) Another important advantage of fiber-optic FMCW interferometers is that they can be combined to form fiber-optic FMCW interferometric networks—the multiplexed fiber-optic interferometers. A multiplexed fiber-optic interferometer usually uses a single light source and a single photodetector, but it can simultaneously measure a number of different targets or different parameters, so that the cost per individual interferometer can be reduced. Another important application of the multiplexed fiber-optic interferometer is that one or more individual interferometers in the network can be used to measure

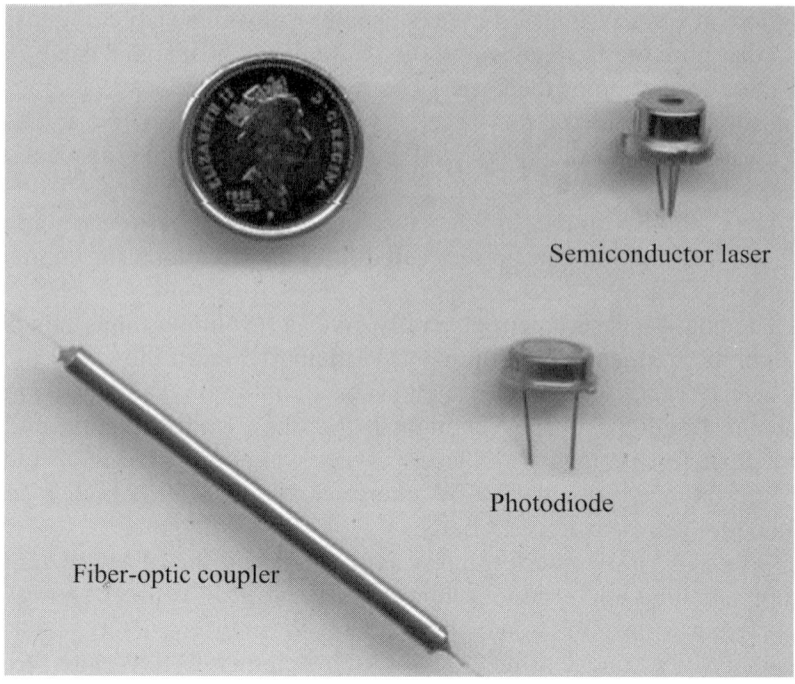

Figure 1.4. Major components of fiber-optic FMCW interferometers. (Photo by the author.)

the effect of environmental conditions on the network (such as temperature), so that the error introduced by the change of environment can be dynamically compensated and the accuracy and long-term stability of the multiplexed inter-ferometer can be significantly improved.

Chapter 2

Principles of Optical Frequency-Modulated Continuous-Wave Interference

In this chapter, we will first review some fundamental concepts of physical optics, then introduce the principles of optical FMCW interference (including sawtooth-wave optical FMCW interference, triangular-wave optical FMCW interference, and sinusoidal-wave optical FMCW interference), and finally discuss multiple-beam optical FMCW interference and multiple-wavelength optical FMCW interference.

2.1 Overview of Physical Optics

First of all, let us recall the conventional theory about optical waves and optical interference. According to electromagnetic theory, an *optical wave* appears to be the propagation of a vibrating electromagnetic field. If the electromagnetic field of a point optical source vibrates at a single frequency and in a fixed direction, the vibrating electric field $E(t)$ can be expressed by

$$
\begin{aligned}
E(t) &= E_0 \cos(\phi) \\
&= E_0 \cos(\omega t + \phi_0) \\
&= E_0 \cos(2\pi \nu t + \phi_0) \\
&= E_0 \cos\left(\frac{2\pi}{T} t + \phi_0\right),
\end{aligned}
$$

$$(2\text{-}1)$$

where E_0 is the vibration amplitude, ϕ is the phase, t is the time, ϕ_0 is the original phase (or initial phase), ω is the vibration angular frequency (or temporal angular frequency), ν is the vibration frequency (or temporal frequency), and T is the vibration period (or temporal period). The parameters ω, ν, and T are related by

$$\omega = 2\pi v$$
$$= \frac{2\pi}{T} .$$

(2-2)

The electric field (or wave function) $E(l,t)$ of the corresponding linearly polarized monochromatic wave at a point in space can be expressed by

$$E(l,t) = E_0(l)\cos\left[\omega\left(t - \frac{l}{v}\right) + \phi_0\right]$$
$$= E_0(l)\cos(\omega t - kl + \phi_0)$$
$$= E_0(l)\cos(2\pi v t - 2\pi fl + \phi_0)$$
$$= E_0(l)\cos\left(\frac{2\pi}{T}t - \frac{2\pi}{\lambda}l + \phi_0\right),$$

(2-3)

where $E_0(l)$ is the wave amplitude whose value is inversely proportional to the propagation distance, l is the distance from the point light source to the point in space under consideration, v is the propagation speed of the wave in the medium, k is the propagation number (or spatial angular frequency), f is the wave number (or spatial frequency), and λ is the wavelength (or spatial period).

The propagation number k is related to the wave number f and the wavelength λ by

$$k = 2\pi f$$
$$= \frac{2\pi}{\lambda} .$$

(2-4)

The temporal and spatial parameters are associated by

$$\lambda = Tv ,$$

(2-5)

$$\omega = kv .$$

(2-6)

Equation (2-3) shows that a linearly polarized monochromatic optical wave is periodic in both space and time. The temporal period is T if the investigation is undertaken along the temporal coordinate, and the spatial period equals λ if the investigation is undertaken along the propagation direction. Figure 2.1 shows the waveforms of a monochromatic optical wave when the investigations are undertaken separately at a fixed space location and at a fixed

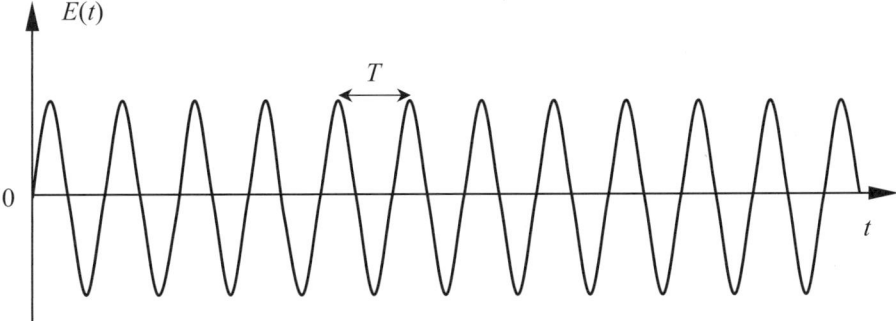

(a) Waveform of the electric field at a fixed location.

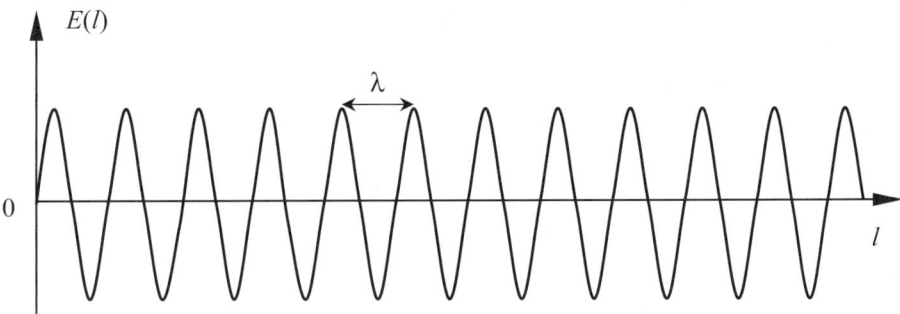

(b) Waveform of the electric field at a fixed moment.

Figure 2.1. Linearly polarized monochromatic optical wave.

time moment.

For convenience, we usually use a complex function to represent the electric field of an optical wave

$$
\begin{aligned}
E(l,t) &= E_0(l)e^{j\left[\omega\left(t-\frac{l}{\upsilon}\right)+\phi_0\right]} \\
&= E_0(l)e^{j(\omega t - kl + \phi_0)} \\
&= E_0(l)e^{j(2\pi\nu t - 2\pi f l + \phi_0)} \\
&= E_0(l)e^{j\left(\frac{2\pi}{T}t - \frac{2\pi}{\lambda}l + \phi_0\right)} .
\end{aligned}
\tag{2-7}
$$

It should be noted that only the real part of the complex expression represents the real physical quantity. Moreover, we often separate the phase term into a spatial term and a temporal term,

$$E(l,t) = E_0(l)e^{-j(kl-\phi_0)}e^{j\omega t}.$$

(2-8)

The amplitude together with the spatial term is called the complex amplitude $E(l)$,

$$E(l) = E_0(l)e^{-j(kl-\phi_0)}.$$

(2-9)

In physical optics, we often deal with the superposition of linearly polarized monochromatic waves of the same frequency. Under this situation, the temporal term is generally omitted because it is a common term for all waves, and only the complex amplitude is taken into consideration. For instance, in an isotropic homogeneous transparent medium, the complex amplitude $E(r)$ of the linearly polarized monochromatic plane wave, which may be produced by a monochromatic point optical source at infinity or at the focal point of a collimating lens, can be expressed by

$$E(r) = E_0 e^{-j(k \cdot r - \phi_0)},$$

(2-10)

where k is the propagation vector and r is the position vector. The propagation vector points to the propagation direction, and its magnitude equals the propagation number k. If the direction of propagation is parallel to the x-axis, the complex amplitude can be simplified as

$$E(x) = E_0 e^{-j(kx - \phi_0)}.$$

(2-11)

The irradiance (or intensity) of an optical wave is defined as the average energy propagated through a unit area in unit time and has been proved to be proportional to (usually represented by) the time average of the square of the electric field magnitude

$$I(r,t) = <E(r,\text{t})^2>,$$

(2-12)

where $<>$ indicates a time average over a period much longer than the vibration period T. If a number of optical waves vibrating in the same direction interfere, the intensity of the resulting electric field will be

$$I(r,t) = <\left[\sum_{i=1}^{m} E_i(r,\text{t})\right]^2>,$$

(2-13)

where i is a positive integer, m is the number of interfering optical waves.

For a linearly polarized monochromatic wave, the intensity is proportional to (usually represented by) the square of the amplitude of the electric field and usually calculated from its complex amplitude

$$I(r) = |E(r)|^2,$$

(2-14)

where $|\ |$ represents the modulus of a complex number. If a number of monochromatic waves of the same frequency interfere, because the resulting electric field is still a monochromatic wave of the same frequency, the intensity of the combined wave can be determined by

$$I(r) = \left| \sum_{i=1}^{m} E_i(r) \right|^2.$$

(2-15)

Obviously, the intensity pattern of this resulting field (also called the interference fringes) is stable in space because it is independent of time. This phenomenon is called *homodyne interference* (or single-frequency interference).

For instance, if two linearly polarized monochromatic plane waves of the same frequency $E_1(r)$ and $E_2(r)$ interfere, the intensity of the resulting electric field can be expressed by

$$I(r) = |E_1(r) + E_2(r)|^2$$

$$= [E_1(r) + E_2(r)][E_1(r) + E_2(r)]^*$$

$$= E_1(r)E_1^*(r) + E_2(r)E_2^*(r) + E_1(r)E_2^*(r) + E_1^*(r)E_2(r)$$

$$= E_{01}^2 + E_{02}^2 + 2E_{01}E_{02}\cos[(k_1 - k_2)\cdot r - (\phi_{01} - \phi_{02})]$$

$$= I_1 + I_2 + 2\sqrt{I_1 I_2}\cos[(k_1 - k_2)\cdot r - (\phi_{01} - \phi_{02})]$$

, (2-16)

where $E_1^*(r)$ and $E_2^*(r)$ are the conjugates of $E_1(r)$ and $E_2(r)$, respectively, I_1 and I_2 are intensities of the two waves, respectively ($I_1 = E_1(r)E_1^*(r) = E_{01}^2$, $I_2 = E_2(r)E_2^*(r) = E_{02}^2$), k_1 and ϕ_{01} are the propagation vector and initial phase of the first wave respectively, and k_2 and ϕ_{02} are the propagation vector and initial phase of the second wave, respectively.

Apparently, the interference effect is represented by the third term. If the initial phases of the two plane waves are random and independent (i.e., *incoherent*), the time average of ($\phi_{01} - \phi_{02}$) will produce a zero coefficient, and there will be no visible interference effect,

$$I(r) = I_1 + I_2.$$

(2-17)

If the initial phases of the two plane waves are correlated (i.e., *coherent*), the intensity of the resulting electric field will be

$$I(\boldsymbol{r}) = I_1 + I_2 + 2\sqrt{I_1 I_2}\,\cos[(\boldsymbol{k}_1 - \boldsymbol{k}_2)\cdot\boldsymbol{r} - (\phi_{01} - \phi_{02})]$$
$$= I_0\{1 + V\cos[(\boldsymbol{k}_1 - \boldsymbol{k}_2)\cdot\boldsymbol{r} - (\phi_{01} - \phi_{02})]\} \qquad , \qquad (2\text{-}18)$$

where I_0 is the average intensity of the resulting electric field given by

$$I_0 = I_1 + I_2 \, , \qquad (2\text{-}19)$$

and V is the visibility (or contrast) of the fringes given by

$$V = \frac{2\sqrt{I_1 I_2}}{I_1 + I_2} \, . \qquad (2\text{-}20)$$

The visibility of the fringes can also be equivalently calculated with the standard formula

$$V = \frac{I_{max} - I_{min}}{I_{max} + I_{min}} \, , \qquad (2\text{-}21)$$

where I_{max} and I_{min} are the maximum and minimum intensities of the resulting electric field, respectively. The visibility of the fringes can have a value between 0 and 1. The maximum visibility will occur when the two waves have equal intensities.

Particularly, if the two plane waves propagate along the same direction, for instance along the x-axis, the intensity of the resulting electric field will be

$$I = I_0[1 + V\cos(\phi_{02} - \phi_{01})] \, . \qquad (2\text{-}22)$$

If these two plane waves are derived from the same light source but traveled along different paths before they meet, the intensity of the resulting electric field will be

$$I(l_1, l_2) = I_0[1 + V\cos(kl_2 - kl_1)]$$
$$= I_0[1 + V\cos(k\Delta l)]$$
$$= I_0[1 + V\cos(k_0 n\Delta l)] \qquad , \qquad (2\text{-}23)$$

where l_1 and l_2 are the geometrical path lengths passed by the two waves, Δl is the geometrical path difference ($\Delta l = l_2 - l_1$), n is the optical refractive index of the medium, and k_0 is the optical propagation number in free space. Usually, the intensity of the resulting electric field is rewritten as

$$\begin{aligned}
I(OPD) &= I_0[1 + V\cos(k_0 n l_2 - k_0 n l_1)] \\
&= I_0[1 + V\cos(k_0 OPL_2 - k_0 OPL_1)] \\
&= I_0[1 + V\cos(k_0 OPD)] \\
&= I_0\left[1 + V\cos\left(\frac{2\pi}{\lambda_0}OPD\right)\right]
\end{aligned}$$

(2-24)

where OPL_1 and OPL_2 are optical path lengths passed by the two waves ($OPL_1 = nl_1$, $OPL_2 = nl_2$), λ_0 is the optical wavelength in free space, and OPD is the optical path difference between the two waves given by

$$\begin{aligned}
OPD &= OPL_2 - OPL_1 \\
&= n\Delta l
\end{aligned}$$

(2-25)

Figure 2.2 shows the waveform of the last expression in Equation (2-24). Obviously, if $OPD = m\lambda_0$ (m is an integer), the intensity of the resulting electric field reaches the maximum.

If two coherent monochromatic plane waves $E_1(r, t)$ and $E_2(r, t)$ are of different frequencies, such as the two waves emitted from a Zeeman laser or obtained by using one or two Bragg acousto-optic modulators operating at different frequencies, the expressions of the electric fields should include both spatial and temporal terms,

$$E_1(r,t) = E_{01}e^{j(\omega_1 t - k_1 \cdot r - \phi_{01})}$$

(2-26)

$$E_2(r,t) = E_{02}e^{j(\omega_2 t - k_2 \cdot r - \phi_{02})}$$

(2-27)

where ω_1 and k_1 are the angular frequency and the propagation vector of the first wave, respectively, and ω_2 and k_2 are the angular frequency and the propagation vector of the second wave, respectively. If the two waves interfere, the resulting electric field will be

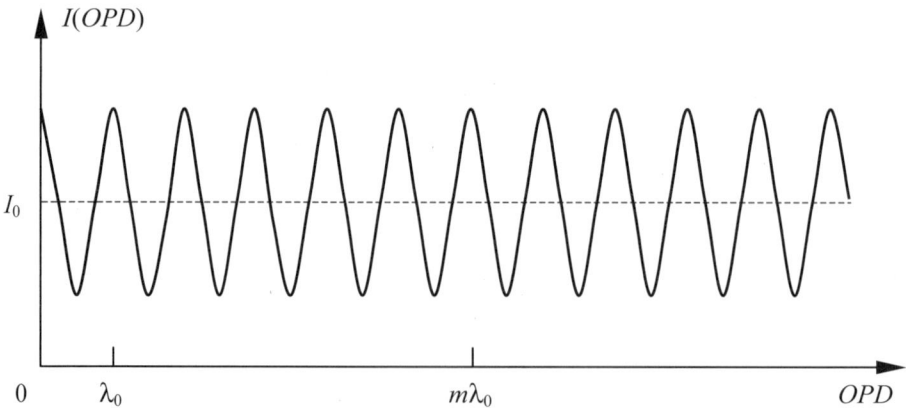

Figure 2.2. Interference fringes of optical homodyne interference.

$$E(r,t) = E_1(r,t) + E_2(r,t)$$

$$= 2E_{01}\cos\left(\frac{\omega_2 - \omega_1}{2}t - \frac{k_2 - k_1}{2}\cdot r + \frac{\phi_{02} - \phi_{01}}{2}\right)e^{j\left(\frac{\omega_2 + \omega_1}{2}t - \frac{k_2 + k_1}{2}\cdot r + \frac{\phi_{02} + \phi_{01}}{2}\right)},$$

$$(2\text{-}28)$$

where we assume $E_{01} = E_{02}$ for simplicity. This combined wave can be treated as a quasi-monochromatic plane wave, whose angular frequency is equal to $(\omega_2+\omega_1)/2$, propagation number is equal to $(k_2+k_1)/2$, initial phase is equal to $(\phi_{02}+\phi_{01})/2$, and amplitude is modulated by a cosine function of a low frequency $(\omega_2-\omega_1)/2$. Therefore, the optical intensity of the resulting electric field should be equal to

$$I(r,t) = \left[2E_{01}\cos\left(\frac{\omega_2 - \omega_1}{2}t - \frac{k_2 - k_1}{2}\cdot r + \frac{\phi_{02} - \phi_{01}}{2}\right)\right]^2$$

$$= 2E_{01}^2\{1 + \cos[(\omega_2 - \omega_1)t - (k_2 - k_1)\cdot r + (\phi_{02} - \phi_{01})]\}$$

$$= I_0\{1 + \cos[(\omega_2 - \omega_1)t - (k_2 - k_1)\cdot r + (\phi_{02} - \phi_{01})]\} \quad , \quad (2\text{-}29)$$

where I_0 is the average intensity of the resulting electric field ($I_0 = 2E_{10}^2$). Obviously, the intensity of the resulting electric field now has a temporal dependence. This phenomenon is called *heterodyne interference* (or double-frequency interference).

If these two plane waves propagate along the same direction (say the *x*-axis), the intensity of the resulting electric field will be

$$I(x,t) = I_0 \{1 + \cos[(\omega_2 - \omega_1)t - (k_2 - k_1)x + (\phi_{02} - \phi_{01})]\} , \qquad (2\text{-}30)$$

where k_1 and k_2 are the propagation numbers of the two waves. Since the intensity of the resulting electric field under this situation is a function of both the spatial and temporal coordinates, no fringe can be seen. However, if a photodetector is placed on the plane perpendicular to the x-axis, a temporal signal (or *beat signal*) can be obtained,

$$
\begin{aligned}
I(x,t) &= I_0[1 + \cos(\omega_b t + \phi_{b0})] \\
&= I_0[1 + \cos(2\pi v_b t + \phi_{b0})] \\
&= I_0\left[1 + \cos\left(\frac{2\pi}{T_b}t + \phi_{b0}\right)\right] ,
\end{aligned}
\qquad (2\text{-}31)
$$

where ω_b is the angular frequency of the beat signal ($\omega_b = \omega_2 - \omega_1$), ϕ_{b0} is the initial phase of the beat signal ($\phi_{b0} = (\phi_{02} - \phi_{01}) - (k_2 - k_1)x$), v_b is the frequency of the beat signal ($v_b = \omega_b / 2\pi$), and T_b is the period of the beat signal ($T_b = 2\pi/\omega_b$), as shown in Figure 2.3.

Notice that the intensity of the combined wave from two different frequency waves can be calculated directly from their complex wave functions:

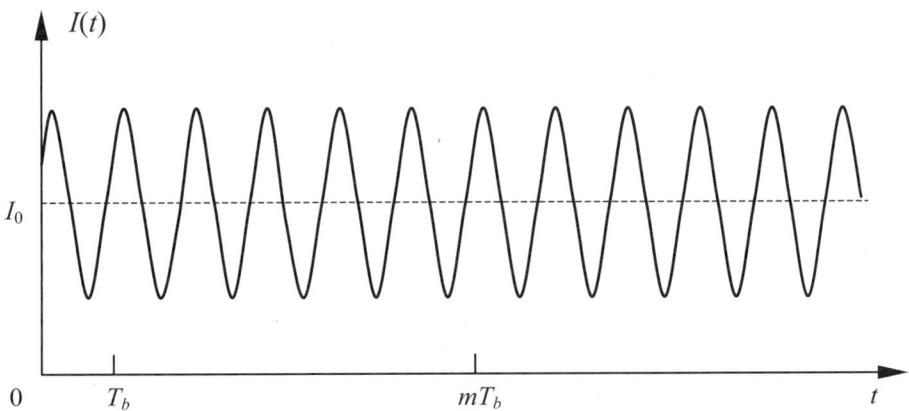

Figure 2.3. Beat signal of optical heterodyne interference.

$$\begin{aligned}
I(\boldsymbol{r},t) &= \left| E(\boldsymbol{r},t) \right| \\
&= [E_1(\boldsymbol{r},t) + E_2(\boldsymbol{r},t)][E_1(\boldsymbol{r},t) + E_2(\boldsymbol{r},t)]^* \\
&= E_1(\boldsymbol{r},t)E_1*(\boldsymbol{r},t) + E_2(\boldsymbol{r},t)E_2*(\boldsymbol{r},t) + E_1(\boldsymbol{r},t)E_2*(\boldsymbol{r},t) \\
&\quad + E_1*(\boldsymbol{r},t)E_2(\boldsymbol{r},t) \\
&= E_{01}^2 + E_{02}^2 + 2E_{01}E_{02}\cos[(\omega_2 - \omega_1)t - (\boldsymbol{k}_2 - \boldsymbol{k}_1)\cdot\boldsymbol{r} + (\phi_{02} - \phi_{01})] \\
&= I_1 + I_2 + 2\sqrt{I_1 + I_2}\cos[(\omega_2 - \omega_1)t - (\boldsymbol{k}_2 - \boldsymbol{k}_1)\cdot\boldsymbol{r} + (\phi_{02} - \phi_{01})] \\
&= I_0\{1 + V\cos[(\omega_2 - \omega_1)t - (\boldsymbol{k}_2 - \boldsymbol{k}_1)\cdot\boldsymbol{r} + (\phi_{02} - \phi_{01})]\}
\end{aligned}$$

.(2-32)

This simple intensity calculation method will also be used in optical FMCW interference, which will be discussed next.

2.2 Optical FMCW Interference[*]

Optical frequency-modulated continuous-wave waves, as the name indicates, are optical waves whose frequencies (or angular frequencies) are continuously modulated. It should be noted that, because the frequency of an optical FMCW wave is not constant, the parameter frequency no longer represents the periodic property of the wave in the conventional sense and therefore must be given another definition.

Whatever the difference, the vibrating electric field of an FMCW light source still includes the amplitude and phase components, but the phase component is a nonlinear function of time. If we define the derivative of the phase as the angular frequency $\omega(t)$

$$\omega(t) = \frac{d\phi(t)}{dt},$$

(2-33)

the phase component $\phi(t)$ can be written as

$$\phi(t) = \int_0^t \omega(t)dt + \phi_0,$$

(2-34)

where ϕ_0 is the initial phase of the light source. The vibrating electric field is

---

*This section is from the author's paper [107]: Jesse Zheng, "Analysis of optical frequency-modulated continuous-wave interference," Applied Optics, Vol. 43, No. 21, pp. 4189–4198, ©2004 Optical Society of America.

$$E(t) = E_0 e^{j\phi(t)},$$ (2-35)

where E_0 is the amplitude of the light source.

Under the situation of small angular frequency modulation ($\Delta\omega << \omega_o$), the chromatic dispersion of the medium can be neglected and therefore the wave function of an optical FMCW wave can be written as

$$E(l,t) = E_0 e^{j\phi\left(t - \frac{l}{\upsilon}\right)},$$ (2-36)

where l is the distance from the light source to the point under consideration and υ is the speed of light in the medium. Usually, the wave function of an optical FMCW wave is rewritten as

$$E(\tau,t) = E_0 e^{j\phi(t-\tau)},$$ (2-37)

where τ is the propagation time of the wave from the light source to the point under consideration, given by

$$\tau = \frac{l}{\upsilon}$$
$$= \frac{nl}{c},$$ (2-38)

where n is the refractive index of the medium and c is the speed of light in free space.

Figure 2.4 shows the waveforms of an optical FMCW wave when the investigation is undertaken separately at a fixed location and at a fixed time moment. Apparently, in both the temporal and special coordinates, different pulses have different periods. In other words, there is no fixed common period or fixed common frequency in both the time and space domains.

As indicated earlier, the intensity of the linearly polarized FMCW wave can still be represented by

$$I = < E(\tau,t)^2 >$$
$$= |E(\tau,t)|^2.$$ (2-39)

If a number of optical FMCW waves are recombined to interfere, the intensity of the resulting electric field (also called the beat signal) will be

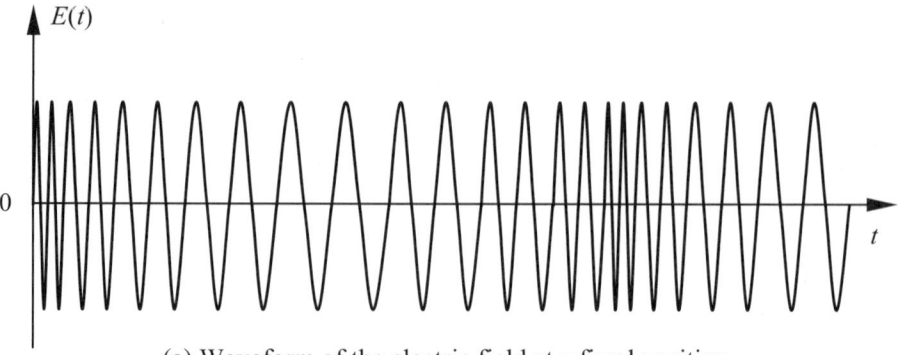

(a) Waveform of the electric field at a fixed position.

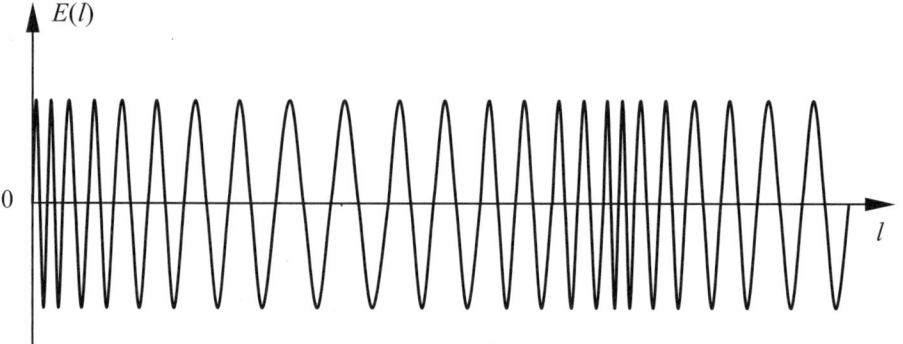

(b) Waveform of the electric field at a fixed moment.

Figure 2.4. Optical frequency-modulated continuous-wave wave.

$$I(\tau_1,...\tau_i,...,t) = \left\langle \left[\sum_{i=1}^{m} E_i(\tau_i,t) \right]^2 \right\rangle$$

$$= \left| \sum_{i=1}^{m} E_i(\tau_i,t) \right|^2$$

(2-40)

For instance, if two linearly polarized optical FMCW plane waves $E_1(\tau_1, t)$ and $E_2(\tau_2, t)$ interfere, the intensity of the resulting electric field can be expressed by

$$I(\tau_1,\tau_2,t) = |E_1(\tau_1,t) + E_2(\tau_2,t)|^2$$
$$= [E_1(\tau_1,t) + E_2(\tau_2,t)][E_1(\tau_1,t) + E_2(\tau_2,t)]^*$$

$$= E_1(\tau_1,t)E_1{}^*(\tau_1,t) + E_2(\tau_2,t)E_2{}^*(\tau_2,t)$$
$$+ E_1(\tau_1,t)E_2{}^*(\tau_2,t) + E_1{}^*(\tau_1,t)E_2(\tau_2,t)$$
$$= E_{01}{}^2 + E_{02}{}^2 + 2E_{01}E_{02}\cos[\phi(t-\tau_1) - \phi(t-\tau_2)]$$
$$= I_1 + I_2 + 2\sqrt{I_1 I_2}\cos[\phi(t-\tau_1) - \phi(t-\tau_2)] \qquad (2\text{-}41)$$

where $E_1{}^*(\tau_1, t)$ and $E_2{}^*(\tau_2, t)$ are the conjugates of $E_1(\tau_1, t)$ and $E_2(\tau_2, t)$, respectively; I_1, E_{01}, and τ_1 are the intensity, amplitude and propagation time of the first wave, respectively ($I_1 = E_1(\tau_1, t)E_1{}^*(\tau_1, t) = E_{01}{}^2$); I_2, E_{02}, and τ_2 are the intensity, amplitude and propagation time of the second wave, respectively ($I_2 = E_2(\tau_2, t)E_2{}^*(\tau_2, t) = E_{02}{}^2$).

Similarly, if the initial phases of the two waves are random and independent, there will be no visible interference effect because the time average of $\phi(t-\tau_1) - \phi(t-\tau_2)$ will produce a zero coefficient. If the initial phases of the two waves are correlated (for instance, the two waves are derived from the same coherent FMCW optical source but travel along different paths before they meet), the intensity of the resulting electric field will be

$$I(\tau_1,\tau_2,t) = I_0\{1 + V\cos[\phi(t-\tau_1) - \phi(t-\tau_2)]\} \qquad (2\text{-}42)$$

where I_0 is the average intensity of the resulting electric field ($I_0 = I_1 + I_2$), V is the contrast of the beat signal ($V = 2\sqrt{I_1 I_2}/(I_1 + I_2)$). Since $\phi(t-\tau_1) - \phi(t-\tau_2) = \phi(t-\tau_1) - \phi[(t-\tau_1) - (\tau_2 - \tau_1)]$, if we shift the origin of the temporal coordinate by τ_1, the intensity of the resulting electric field will become

$$I(\tau,t) = I_0\{1 + V\cos[\phi(t) - \phi(t-\tau)]\} \qquad (2\text{-}43)$$

where τ is the delay time of the second wave (usually called the signal wave) with respect to the first wave (usually called the reference wave), given by

$$\tau = \tau_2 - \tau_1 . \qquad (2\text{-}44)$$

The similarity between optical FMCW interference and optical heterodyne interference is that the intensity of the resulting electric field is dependent on both the spatial and temporal coordinates. To obtain a usable signal, a photodetector that converts the optical signal to an electric signal is required. Moreover, since the sensing surface of the real photodetector has a certain area and shape (normally a plane), to acquire a signal with the best contrast, a suitable optical arrangement is also needed. Therefore, in practice, we always make the optical FMCW waves collimated and parallel when they interfere and place a photodetector on the plane perpendicular to the propagation direction of the waves to detect the beat signal.

The difference between optical FMCW interference and optical heterodyne interference is that the optical FMCW beat signal generally is more complex than the optical heterodyne beat signal. The former is a complicated periodic function, but the latter is a sinusoidal function of a single frequency. On the other hand, the optical FMCW beat signal contains more information than the optical heterodyne beat signal because the frequency of the optical FMCW beat signal is also related to the delay time or optical path difference.

The waveform used for modulating the frequency (or angular frequency) of an optical FMCW wave can be of various shapes, but it should be a periodic function so that the frequency of the optical wave can be continually modulated. Other constraints for the frequency modulation waveform include that the waveform should be easy to generate and that the beat signal produced should be simple to process. In the following subsections, we will discuss the optical FMCW interference with three different waveforms: sawtooth wave, triangular wave, and sinusoidal wave.

2.2.1 Sawtooth-Wave Optical FMCW Interference

If two optical waves derived from the same coherent light source, whose angular frequency is modulated with a sawtooth waveform, but traveled along different paths are recombined at a point in space, they will interfere. The angular frequency waveforms of the two interfering waves and the beat signal produced can be described by the curves in Figure 2.5, where the solid curve represents the angular frequency of the reference wave, the dashed curve stands for the angular frequency of the signal wave, and the dot-dashed curve corresponds to the angular frequency of the beat signal produced.

The angular frequency of the reference wave $\omega_1(t)$ in the period $(-T_m/2 + \tau, T_m/2)$ can be written as

$$\omega_1(t) = \alpha t + \omega_0 ,$$

(2-45)

where ω_0 is the angular frequency at the center of the modulation period (or called the central angular frequency), α is the angular frequency modulation rate, which is given by

$$\alpha = \frac{\Delta\omega}{T_m} ,$$

(2-46)

where $\Delta\omega$ is the angular frequency modulation excursion, and T_m is the period of the modulation signal (or modulation period). The phase of the reference

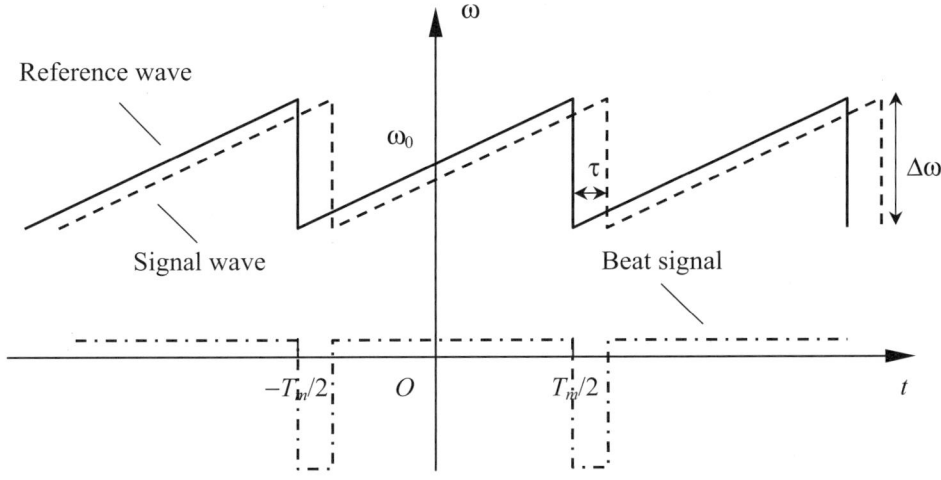

Figure 2.5. Angular frequency relationships between the interfering waves and the beat signal produced in sawtooth-wave FMCW interference.

wave $\phi_1(t)$ can be written as

$$\phi_1(t) = \frac{1}{2}\alpha t^2 + \omega_0 t + \phi_0 ,$$

(2-47)

where ϕ_0 is the initial phase of the light source. The wave function of the reference wave $E_1(t)$ can be written as

$$E_1(t) = E_{01}e^{j\left(\frac{1}{2}\alpha t^2 + \omega_0 t + \phi_0\right)},$$

(2-48)

where E_{01} is the amplitude of the reference wave.

Similarly, for the signal wave, the angular frequency $\omega_2(t)$, phase $\phi_2(t)$ and wave function $E_2(\tau, t)$ can be written as

$$\omega_2(\tau, t) = \alpha(t - \tau) + \omega_0 ,$$

(2-49)

$$\phi_2(\tau, t) = \frac{1}{2}\alpha(t - \tau)^2 + \omega_0(t - \tau) + \phi_0 ,$$

(2-50)

$$E_2(\tau, t) = E_{02}e^{j\left[\frac{1}{2}\alpha(t-\tau)^2 + \omega_0(t-\tau) + \phi_0\right]},$$

(2-51)

where E_{02} is the amplitude of the signal wave, and τ is the delay time of the signal wave with respect to the reference wave.

When these two waves interfere, the intensity $I(\tau, t)$ of the resulting electric field (i.e., the beat signal) is given by

$$
\begin{aligned}
I(\tau,t) &= \left| E_1(t) + E_2(\tau,t) \right|^2 \\
&= [E_1(t) + E_2(\tau,t)][E_1(t) + E_2(\tau,t)]^* \\
&= E_1(t)E_1{}^*(t) + E_2(\tau,t)E_2{}^*(\tau,t) + E_1(t)E_2{}^*(\tau,t) + E_1{}^*(t)E_2(\tau,t) \\
&= E_{01}{}^2 + E_{02}{}^2 + 2E_{01}E_{02}\cos\left(\alpha\tau t + \omega_0\tau - \frac{\alpha\tau^2}{2}\right) \\
&= I_1 + I_2 + 2\sqrt{I_1 I_2}\cos\left(\alpha\tau t + \omega_0\tau - \frac{\alpha\tau^2}{2}\right) \\
&= I_0\left[1 + V\cos\left(\alpha\tau t + \omega_0\tau - \frac{\alpha\tau^2}{2}\right)\right]
\end{aligned}
$$

,(2-52)

where I_1 and I_2 are the intensities of the reference wave and signal wave, respectively ($I_1 = E_{01}{}^2$, $I_2 = E_{02}{}^2$), I_0 is the average intensity of the beat signal ($I_0 = I_1 + I_2$), and V is the contrast of the beat signal ($V = 2\sqrt{I_1 I_2}/(I_1 + I_2)$).

In practice, due to the limitation of the coherence length and frequency modulation range of the available FMCW light sources (for example, for the commercial single-mode semiconductor laser, the coherence length is about 10 meters and the frequency modulation range is less than 100 GHz), $\tau < 3 \times 10^{-8}$ sec, and $\alpha \approx \omega_0$. Therefore, the second-order small quantity ($\alpha\tau^2/2$) in this equation can be neglected (usually treated as a system error), and the beat signal can be simplified as

$$
\begin{aligned}
I(\tau,t) &= I_0[1 + V\cos(\alpha\tau t + \omega_0\tau)] \\
&= I_0[1 + V\cos(\omega_b t + \phi_{b0})]
\end{aligned}
$$

, (2-53)

where ω_b is the angular frequency of the beat signal given by

$$
\omega_b = \alpha\tau
$$

, (2-54)

and ϕ_{b0} is the initial phase of the beat signal given by

$$
\phi_{b0} = \omega_0\tau
$$

. (2-55)

In the period $(-T_m/2, -T_m/2 +\tau)$, the angular frequency, phase, and wave function of the reference wave can still be described by Equations (2-45), (2-47), and (2-48), respectively. However, the corresponding parameters of the signal wave will be

$$\omega_2(\tau,t) = \alpha[t - (\tau - T_m)] + \omega_0 ,$$
(2-56)

$$\phi_2(\tau,t) = \frac{1}{2}\alpha[t - (\tau - T_m)]^2 + \omega_0[t - (\tau - T_m)] + \phi_0 ,$$
(2-57)

$$E_2(\tau,t) = E_{02}e^{j\left\{\frac{1}{2}\alpha[t-(\tau-T_m)]^2 + \omega_0[t-(\tau-T_m)]+\phi_0\right\}} .$$
(2-58)

The beat signal in the period $(-T_m/2, \tau - T_m/2)$ will be

$$I(\tau,t) = I_0\left\{1 + V\cos\left[\alpha(\tau - T_m)t + \omega_0(\tau - T_m) - \frac{1}{2}\alpha(\tau - T_m)^2\right]\right\} .$$
(2-59)

The angular frequency of the beat signal in the period $(-T_m/2, -T_m/2+\tau)$ equals

$$\omega_b = \alpha(\tau - T_m) .$$
(2-60)

Since, in practice, $\tau \le 3\times10^{-8}$ sec, $\omega_b \approx \Delta\omega = 1\times10^{11}$ rad·Hz, the available photodetectors cannot identify such a high frequency in such a short period. Instead, they can only give an average intensity in the period $(-T_m/2, -T_m/2 +\tau)$. Therefore, the beat signal in the full modulation period $(-T_m/2, T_m/2)$ can be approximately represented by

$$I(\tau,t) = I_0[1 + V\cos(\alpha\tau t + \omega_0\tau)]$$
$$= I_0[1 + V\cos(\omega_b t + \phi_{b0})] ,$$
(2-53)

The beat signal in the entire time domain $(-\infty, \infty)$ can be expressed by

$$I(\tau,t) = I_0\{1 + [V\cos(\alpha\tau t + \omega_0\tau)\,\text{rect}_{T_m}(t) \otimes \sum_{m=-\infty}^{\infty}\delta(t - mT_m)]\}$$

$$= I_0\{1 + [V\cos(\omega_b t + \phi_{b0})\,\text{rect}_{T_m}(t) \otimes \sum_{m=-\infty}^{\infty}\delta(t - mT_m)]\} ,$$
(2-61)

where $\otimes$ is the convolution operator, m is an integer, and the rect_T function is defined as

$$\mathrm{rect}_T(t) = \begin{cases} 1 & |t| \le T/2 \\ 0 & |t| > T/2 \end{cases}. \tag{2-62}$$

In physical optics, we often use frequency, wavelength, and optical path difference to describe optical waves and optical interference phenomena. Thus, the beat signal in each modulation period can be rewritten as

$$\begin{aligned} I(OPD, t) &= I_0\left[1 + V\cos\left(\frac{2\pi\Delta v v_m OPD}{c}t + \frac{2\pi}{\lambda_0}OPD\right)\right] \\ &= I_0[1 + V\cos(2\pi v_b t + \phi_{b0})] \end{aligned}, \tag{2-63}$$

where Δv is the optical frequency modulation excursion ($\Delta v = \Delta\omega/2\pi$), v_m is the frequency of the modulation signal (or modulation frequency) ($v_m = 1/T_m$), OPD is the optical path difference ($OPD = c\tau$), c is the speed of light in free space, λ_0 is the central optical wavelength in free space ($\lambda_0 = 2\pi c/\omega_0$), and v_b and ϕ_{b0} are the frequency and the initial phase of the beat signal

$$\begin{aligned} v_b &= \frac{\alpha OPD}{2\pi c} \\ &= \frac{\Delta v v_m OPD}{c} \end{aligned}, \tag{2-64}$$

$$\begin{aligned} \phi_{b0} &= \frac{\omega_0 OPD}{c} \\ &= k_0 OPD \\ &= \frac{2\pi OPD}{\lambda_0} \end{aligned}, \tag{2-65}$$

where k_0 is the central propagation number in free space ($k_0 = \omega_0/c$). Finally, the intensity of the beat signal in the entire time domain can be rewritten as

$$\begin{aligned} I(OPD, t) &= I_0\left\{1 + \left[V\cos\left(\frac{2\pi\Delta v v_m OPD}{c}t + \frac{2\pi}{\lambda_0}OPD\right)\mathrm{rect}_{T_m}(t) \otimes \sum_{m=-\infty}^{\infty}\delta(t - mT_m)\right]\right\} \\ &= I_0\{1 + [V\cos(\omega_b t + \phi_{b0})\,\mathrm{rect}_{T_m}(t) \otimes \sum_{m=-\infty}^{\infty}\delta(t - mT_m)]\} \end{aligned}$$

$$\tag{2-66}$$

Equation (2-65) shows that, if the *OPD* changes by one wavelength, the beat signal will shift by one period. This sounds the same as the signal of optical homodyne interference, but actually they are completely different. The optical FMCW beat signal is a dynamic signal (i.e., a continuous function of the time), and therefore, to calibrate the fractional phase, distinguish the shift direction, and count the number of the full periods is much easier than for the static fringes. Also different from the beat signal of optical heterodyne interference is the fact that, the angular frequency of the FMCW beat signal is proportional to the *OPD*, and can therefore be used to determine the absolute value of the *OPD*.

Figure 2.6 shows the waveform of the beat signal produced by two coherent sawtooth-wave optical FMCW waves. Note that the phase of the beat signal usually is not continuous at the junctions of the modulation periods unless the beat frequency is coincidentally equal to a multiple of the modulation frequency. This implies that the beat signal generally is not a regular sinusoidal wave.

The photo in Figure 2.7 shows the real signals from the Mach-Zehnder FMCW interferometer illuminated by a single-mode semiconductor laser whose frequency is modulated with a sawtooth waveform (see Section 6.3). The upper trace is the waveform of the drive current for the semiconductor laser representing the optical frequency, and the lower trace is the waveform of the beat signal produced. A small variation in the amplitude of the beat signal is due to the associated intensity modulation of the semiconductor laser.

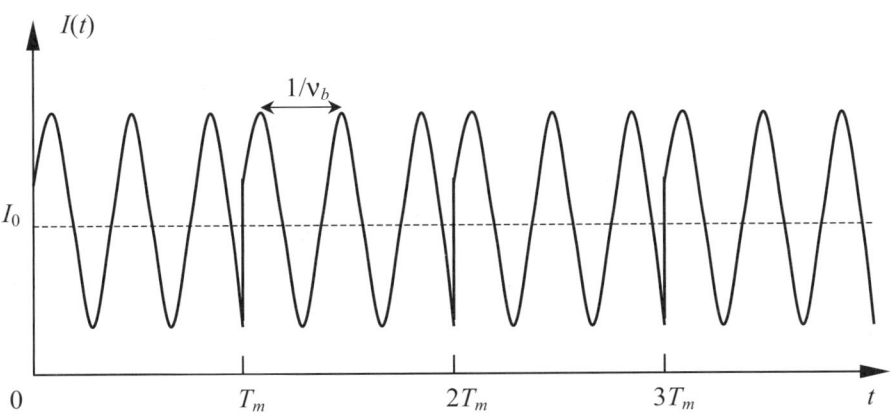

Figure 2.6. Waveform of the beat signal produced by two coherent sawtooth-wave FMCW waves.

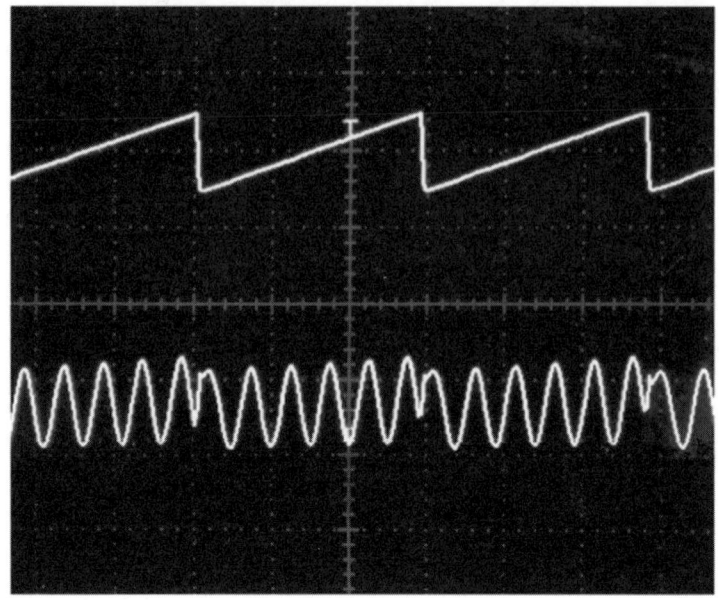

Figure 2.7. Waveforms of signals from a real optical sawtooth-wave FMCW interferometer. The upper trace is the waveform of the frequency modulation signal; the lower trace is the waveform of the beat signal. (Photo by the author.)

Up to this point, the delay time between the two interfering waves has been assumed to be constant. If the delay time $\tau(t)$ varies with time, the detected signal $I'(\tau, t)$ in each modulation period will be

$$I'(\tau,t) = I_0\{1 + V\cos[\alpha\tau(t)t + \omega_0\tau(t)]\}$$. (2-67)

The angular frequency ω_b' of the detected signal will be

$$\omega_b' = \frac{d}{dt}[\alpha\tau(t)t + \omega_0\tau(t)]$$

$$= \alpha\tau(t) + (\omega_0 + \alpha t)\frac{d\tau(t)}{dt}$$

$$= \omega_b + \omega_D$$, (2-68)

where ω_b is the real beat angular frequency, and ω_D is the Doppler angular frequency shift given by

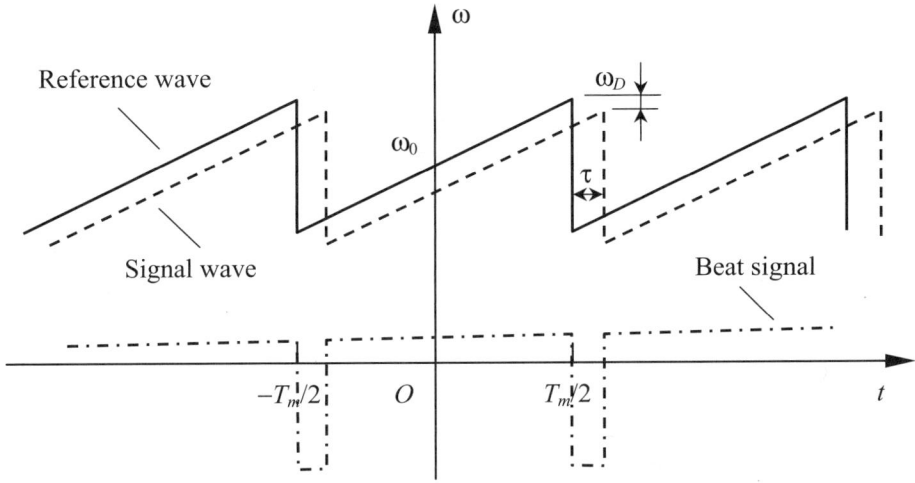

Figure 2.8. Angular frequency relationships between the interfering waves and the detected signal in sawtooth-wave FMCW interference when the delay time is changing.

$$\omega_D = (\omega_0 + \alpha t)\frac{d\tau(t)}{dt} .$$

(2-69)

Notice that the Doppler angular frequency shift is a function of time, but its average value $\overline{\omega_D}$ is equal to

$$\overline{\omega_D} = \omega_0 \frac{d\tau(t)}{dt} .$$

(2-70)

Equation (2-68) shows that the angular frequency of the detected signal contains two parts: one is the contribution of the optical FMCW interference, which is related to the delay time; the other is the contribution of the Doppler effect, which is related to the derivative of the delay time. Figure 2.8 shows the angular frequency waveforms of the two interfering waves and the detected signal when the delay time is changing.

Similarly, in terms of frequency, wavelength, and optical path difference, the average Doppler frequency shift $\overline{v_D}$ can be written as

$$\overline{v_D} = \frac{1}{\lambda_0}\frac{dOPD(t)}{dt} ,$$

(2-71)

where λ_0 is the central wavelength in free space. If the signal wave is a reflective wave reflected from a moving object, the average Doppler frequency shift will be

$$\overline{v_D} = \frac{2n}{\lambda_0} s ,$$

(2-72)

where s is the speed of the moving object, and n is the refractive index of the medium.

The Doppler effect introduces an error in measurement of the beat frequency, and therefore sawtooth-wave FMCW interference is only suitable for measuring the static or slow-changing *OPD*.

2.2.2 Triangular-Wave Optical FMCW Interference

For convenience, let's define the period of the triangular-wave modulation signal as $2T_m$ and divide each period into two parts, the rising period and the falling period, as shown in Figure 2.9. When two coherent triangular-wave FMCW waves interfere, in the rising period, the intensity of the resulting electric field $I_r(\tau, t)$, beat frequency ω_{br}, and initial phase ϕ_{b0r} are the same as those in sawtooth-wave FMCW interference,

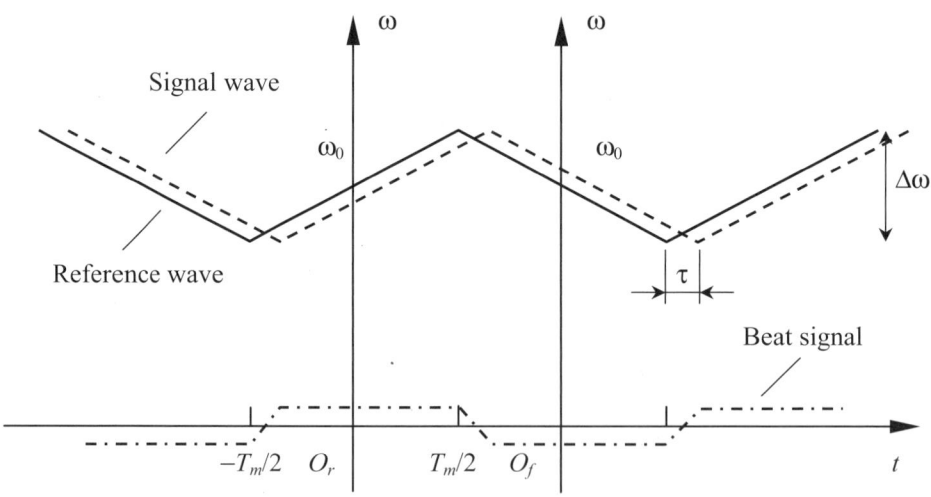

Figure 2.9. Angular frequency relationships between the interfering waves and the beat signal in triangular-wave FMCW interference.

$$I_r(\tau, t) = I_0[1 + V\cos(\alpha\tau t + \omega_0\tau)],$$ (2-73)

$$\omega_{br} = \alpha\tau,$$ (2-74)

$$\phi_{b0r} = \omega_0\tau,$$ (2-75)

where α is the angular frequency modulation rate in the rising period ($\alpha \geq 0$) and ω_0 is the angular frequency at the center of the rising period. In the falling period, however, because the angular frequency modulation rate is negative, the intensity of the resulting electric field $I_f(\tau, t)$, beat frequency ω_{bf}, and initial phase ϕ_{b0f} become

$$I_f(\tau, t) = I_0[1 + V\cos(-\alpha\tau t + \omega_0\tau)],$$ (2-76)

$$\omega_{bf} = -\alpha\tau,$$ (2-77)

$$\phi_{b0f} = \omega_0\tau.$$ (2-78)

Note that Equations (2-73) and (2-76) are based on different origins in the temporal coordinate axis.

In general, we cannot tell the sign of an angular frequency; instead, we can detect only the absolute value of an angular frequency. Therefore, measuring the angular frequency of the beat signal in either the rising period or the falling period, we can find out the delay time between the two interfering waves ($|\omega_{br}| = |\omega_{bf}| = \alpha\tau$).

Figure 2.10 shows the waveform of the beat signal produced by the two coherent triangular-wave FMCW waves. Notice that the phase of the beat signal is always continuous at the junctions of the periods, but the phase shift directions in the rising and falling periods are opposite when the delay time is changing. The photo in Figure 2.11 shows the waveforms of the real signals from the Mach-Zehnder FMCW interferometer illuminated by a single-mode semiconductor laser whose frequency is modulated with a triangular wave. The upper trace is the waveform of the drive current for the semiconductor laser; the lower trace is the waveform of the beat signal produced.

If the delay time is changing, from Equations (2-73) and (2-76), we can find that the average angular frequency of the detected signal in the rising period $\overline{\omega_{br}}'$ equals

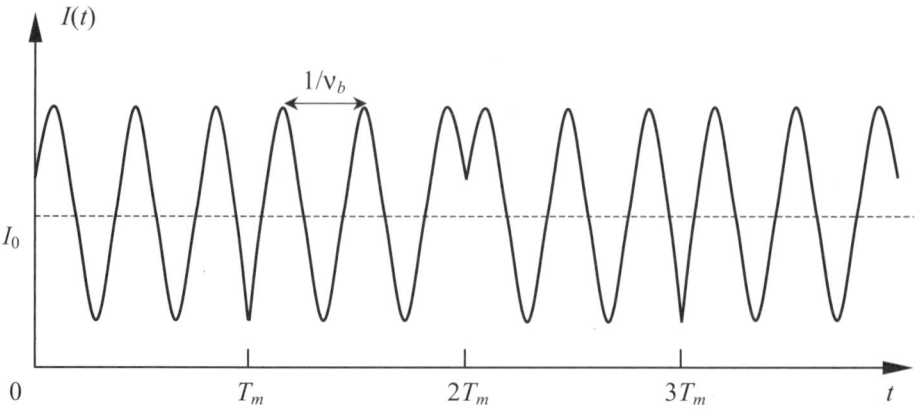

Figure 2.10. Waveform of the beat signal produced by two coherent triangular-wave FMCW waves.

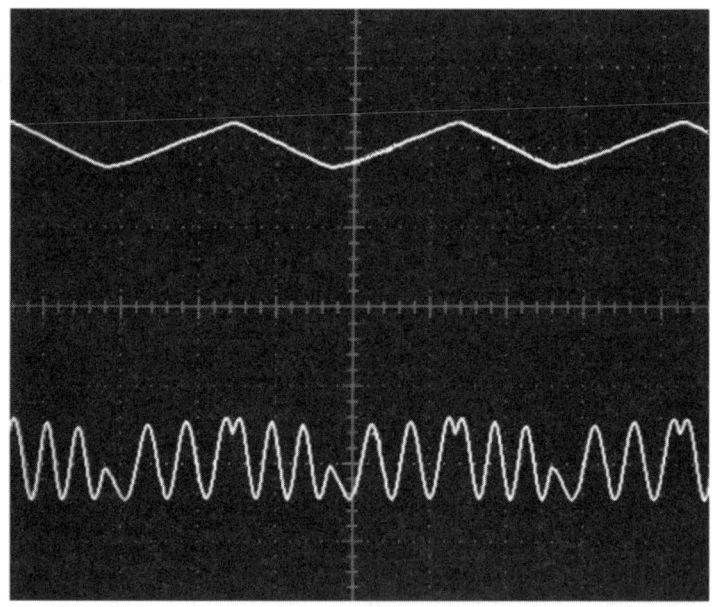

Figure 2.11. Waveforms of signals from a real optical triangular-wave FMCW interferometer. The upper trace is the waveform of the frequency modulation signal; the lower trace is the waveform of the beat signal produced. (Photo by the author.)

$$\overline{\omega_{br}'} = \overline{\omega_b} + \overline{\omega_D}, \tag{2-79}$$

where $\overline{\omega_D}$ is the average Doppler angular frequency shift $\left(\overline{\omega_D} = \omega_0 \dfrac{d\tau(t)}{dt} \right)$, and that the angular frequency of the detected signal in the falling period $\overline{\omega_{bf}'}$ equals

$$\overline{\omega_{bf}'} = \overline{\omega_D} - \overline{\omega_b}, \tag{2-80}$$

Figure 2.12 illustrates the angular frequency waveforms of the two interfering triangular-wave FMCW waves and the detected signal when the delay time is changing.

Because $\overline{\omega_{bf}'}$ is negative, Equation (2-80) can be rewritten as

$$\left| \overline{\omega_{bf}'} \right| = \overline{\omega_b} - \overline{\omega_D}. \tag{2-81}$$

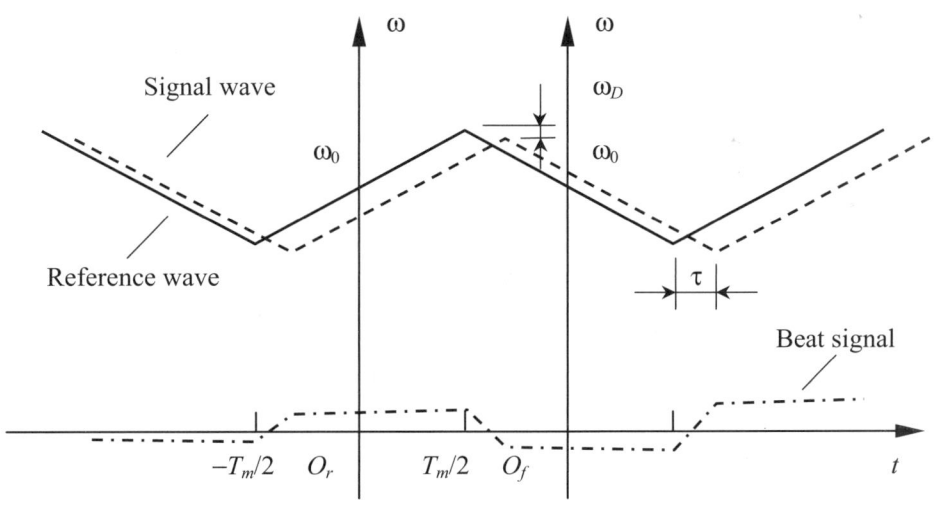

Figure 2.12. Angular frequency relationship between the interfering waves and the beat signal in triangular-wave FMCW interference when the delay time is changing.

Evidently, the average value of $\overline{\omega_{br}'}$ and $\overline{|\omega_{bf}'|}$ is equal to the angular frequency of the real beat signal

$$\omega_b = \frac{1}{2}\left(\overline{\omega_{br}'} + \overline{|\omega_{bf}'|}\right).$$

(2-82)

Half the difference between $\overline{\omega_{br}'}$ and $\overline{|\omega_{bf}'|}$ is equal to the average Doppler angular frequency shift

$$\overline{\omega_D} = \frac{1}{2}\left(\overline{\omega_{br}'} - \overline{|\omega_{bf}'|}\right).$$

(2-83)

Hence, if we measure the average angular frequencies of the detected signal in the rising period and the falling period separately, we can calculate the real beat angular frequency by using Equation (2-82) and compute the true average Doppler angular frequency shift by using Equation (2-83). Consequently, we can find the correct value of the delay time or *OPD* and the speed of a moving object.

For instance, if the signal wave is a reflective wave reflected from a moving object, based on Equation (2-72), we can find that the speed of the moving object *s* will be

$$s = \frac{\lambda_0}{2n}\overline{v_D},$$

(2-84)

where λ_0 is the central wavelength in free space, n is the refractive index of the medium, and $\overline{v_D}$ is the average Doppler frequency shift.

Sawtooth-wave FMCW interference and triangular-wave FMCW interference belong to linear FMCW interference. The advantages of linear FMCW interference include single beat frequency and linear response to the delay time or *OPD*. The drawback of linear FMCW interference is that it requires a linear frequency modulation light source, which generally is relatively more difficult to achieve than a sinusoidal-wave frequency modulation light source.

2.2.3 Sinusoidal-Wave Optical FMCW Interference

In sinusoidal-wave FMCW interference, the angular frequency of the reference wave $\omega_1(t)$ can be expressed by

$$\omega_1(t) = \omega_0 + \frac{\Delta\omega}{2}\sin(\omega_m t),$$

(2-85)

where ω_0 is the average angular frequency, $\Delta\omega$ is the peak-to-peak angular frequency modulation excursion, and ω_m is the angular frequency of the modulation signal. The phase of the reference wave $\phi_1(t)$ can be represented by

$$\phi_1(t) = \omega_0 t - \frac{\Delta\omega}{2\omega_m}\cos(\omega_m t) + \phi_0 \quad,$$

(2-86)

where ϕ_0 is the initial phase of the light source. The wave function $E_1(t)$ of the reference wave can be written as

$$E_1(t) = E_{01}e^{j\left[\omega_0 t - \frac{\Delta\omega}{2\omega_m}\cos(\omega_m t) + \phi_0\right]} \quad,$$

(2-87)

where E_{01} is the amplitude.

Similarly, the angular frequency $\omega_2(\tau, t)$, phase $\phi_2(\tau, t)$ and wave function $E_2(\tau, t)$ of the signal wave can be written as

$$\omega_2(\tau, t) = \omega_0 + \frac{\Delta\omega}{2}\sin[\omega_m(t - \tau)] \quad,$$

(2-88)

$$\phi_2(\tau, t) = \omega_0(t - \tau) - \frac{\Delta\omega}{2\omega_m}\cos[\omega_m(t - \tau)] + \phi_0 \quad,$$

(2-89)

$$E_2(\tau, t) = E_{02}e^{j\left\{\omega_0(t-\tau) - \frac{\Delta\omega}{2\omega_m}\cos[\omega_m(t-\tau)] + \phi_0\right\}} \quad,$$

(2-90)

where τ is the delay time of the signal wave with respect to the reference wave.

Figure 2.13 illustrates the angular frequency waveforms of the two interfering sinusoidal-wave FMCW waves and the angular frequency waveform of the beat signal produced, where the solid curve represents the angular frequency of the reference wave, the dashed curve stands for the angular frequency of the signal wave, and the dot-dashed curve corresponds to the angular frequency of the beat signal produced.

When these two waves interfere, the intensity of the resulting electric field $I(\tau, t)$ would be

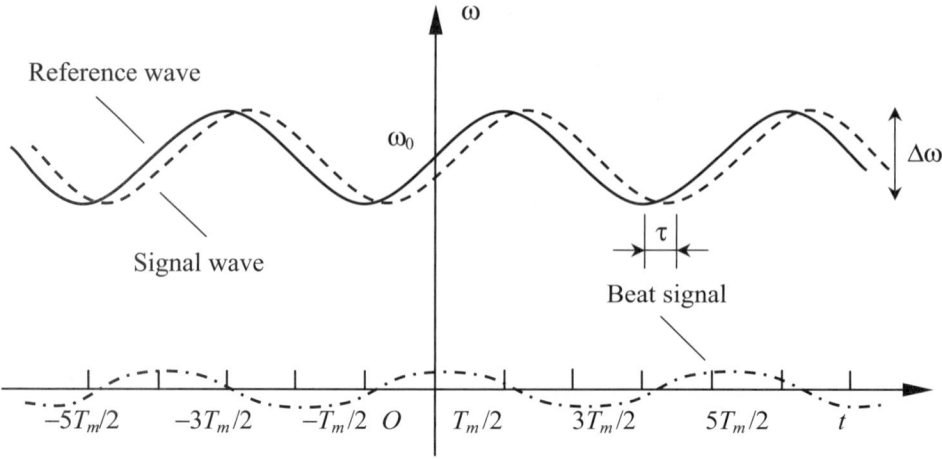

Figure 2.13. Angular frequency relationship between the interfering waves and the beat signal produced in sinusoidal-wave FMCW interference.

$$
\begin{aligned}
I(\tau,t) &= |\, E_1(t) + E_2(\tau,t)\,|^2 \\
&= [E_1(t) + E_2(\tau,t)][E_1(t) + E_2(\tau,t)]\,^* \\
&= E_1(t)E_1{}^*(t) + E_2(\tau,t)E_2{}^*(\tau,t) + E_1(t)E_2{}^*(\tau,t) + E_1{}^*(t)E_2(\tau,t) \\
&= E_{01}{}^2 + E_{02}{}^2 + 2E_{01}E_{02}\cos\left\{\frac{\Delta\omega}{2\omega_m}[\cos\omega_m(t-\tau) - \cos\omega_m t] + \omega_0\tau\right\} \\
&= I_1 + I_2 + 2\sqrt{I_1 I_2}\cos\left\{\frac{\Delta\omega}{2\omega_m}[\cos\omega_m(t-\tau) - \cos\omega_m t] + \omega_0\tau\right\}
\end{aligned}
$$

$$(2\text{-}91)$$

where I_1 is the intensity of the reference wave ($I_1 = E_{01}{}^2$), and I_2 is the intensity of the signal wave ($I_2 = E_{02}{}^2$). This equation can be rewritten as

$$
I(\tau,t) = I_0\left\{1 + V\cos\left[\frac{\Delta\omega}{\omega_m}\sin\left(\frac{\omega_m\tau}{2}\right)\sin\omega_m\left(t - \frac{1}{2}\tau\right) + \omega_0\tau\right]\right\},
$$

$$(2\text{-}92)$$

where I_0 is the average intensity of the beat signal ($I_0 = I_1 + I_2$), and V is the contrast of the beat signal ($V = 2\sqrt{I_1 I_2}\,/(I_1 + I_2)$).

Similarly, considering the real situation, $\tau \le 3\times10^{-8}$ sec, $\omega_m \le 1\times10^5$ rad·Hz, $\omega_m\tau \approx 3\times10^{-3} \ll 1$, $\sin\dfrac{\omega_m\tau}{2} \approx \dfrac{\omega_m\tau}{2}$, $\sin\left[\omega_m\left(t-\dfrac{1}{2}\tau\right)\right] \approx \sin(\omega_m t)$, and therefore,

$$I(\tau,t) = I_0\left\{1 + V\cos\left[\frac{\Delta\omega\tau}{2}\sin(\omega_m t) + \omega_0\tau\right]\right\}.$$

(2-93)

The angular frequency of the beat signal can be expressed by

$$\omega_b = \frac{\Delta\omega\omega_m\tau}{2}\cos(\omega_m t),$$

(2-94)

and the initial phase of the beat signal can be written as

$$\phi_{b0} = \omega_0\tau.$$

(2-95)

Equation (2-94) shows that the angular frequency of the beat signal in sinusoidal-wave FMCW interference is not constant. However, the average absolute value $\overline{|\omega_b|}$ of the angular frequency in a modulation period is equal to

$$\overline{|\omega_b|} = \frac{\Delta\omega\omega_m\tau}{\pi}.$$

(2-96)

If the period of the sinusoidal wave is defined as $2T_m$, this equation becomes

$$\overline{|\omega_b|} = \frac{\Delta\omega\tau}{T_m}$$

$$= \alpha\tau,$$

(2-97)

where α is the effective angular frequency modulation rate ($\alpha = \Delta\omega/T_m$). Hence, using the average value of the beat frequency still yields the delay time or *OPD*.

Figure 2.14 shows the waveform of the beat signal produced by two coherent sinusoidal-wave FMCW waves. The photo in Figure 2.15 shows the real signals from an optical sinusoidal-wave FMCW interferometer equipped with a single-mode semiconductor laser. The upper trace is the waveform of the driving current for the semiconductor laser; the lower trace is the waveform of the beat signal produced.

Similarly, if the delay time is changing, the average angular frequency of the detected signal will contain two parts, the average beat angular frequency and the

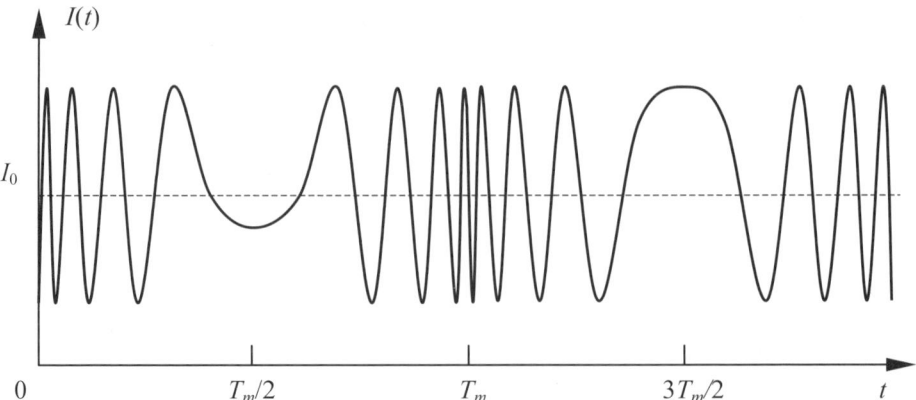

Figure 2.14. Waveform of the beat signal produced by two coherent sinusoidal-wave FMCW waves.

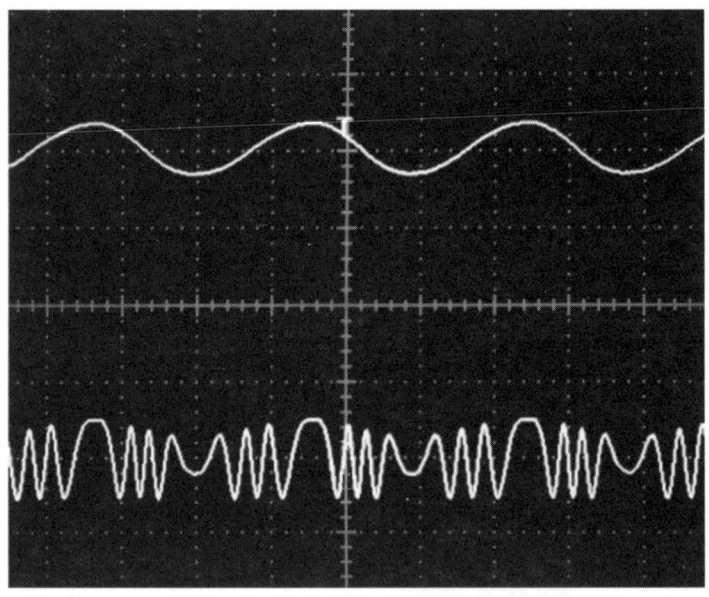

Figure 2.15. Beat signal produced by two coherent optical sinusoidal-wave FMCW waves. The upper trace is the waveform of the frequency modulation signal; the lower trace is the waveform of the beat signal produced. (Photo by the author.)

average Doppler angular frequency shift, as shown in Figure 2.16. In the rising period $(-T_m/2, T_m/2)$, the average angular frequency of the detected signal $\overline{\omega_{br}}'$ equals

$$\overline{\omega_{br}}' = \overline{\omega_b} + \overline{\omega_D} , \tag{2-98}$$

where $\overline{\omega_b}$ is the average real beat angular frequency, and $\overline{\omega_D}$ is the average Doppler angular frequency shift $\left(\overline{\omega_D} = \omega_0 \dfrac{d\tau(t)}{dt}\right)$. In the falling period $(T_m/2, 3T_m/2)$, the absolute value of the average angular frequency of the detected signal $\left|\overline{\omega_{bf}}'\right|$ equals

$$\left|\overline{\omega_{bf}}'\right| = \overline{\omega_b} - \overline{\omega_D} . \tag{2-99}$$

The average value of $\overline{\omega_{br}}'$ and $\left|\overline{\omega_{bf}}'\right|$ determines the average real beat angular frequency

$$\overline{\omega_b} = \frac{1}{2}\left(\overline{\omega_{br}}' + \left|\overline{\omega_{bf}}'\right|\right) . \tag{2-100}$$

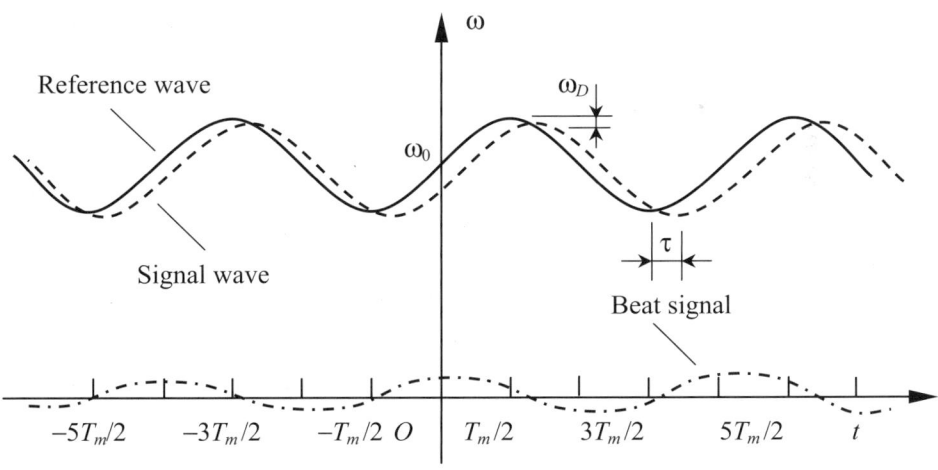

Figure 2.16. Angular frequency relationship between the interfering waves and the beat signal produced in sinusoidal-wave FMCW interference when the delay time is changing.

Half the difference between $\overline{\omega_{br}}'$ and $\left|\overline{\omega_{bf}}'\right|$ gives the average Doppler angular frequency shift

$$\overline{\omega_D} = \frac{1}{2}\left(\overline{\omega_{br}}' - \left|\overline{\omega_{bf}}'\right|\right).$$

(2-101)

Hence, based on the average angular frequencies of the detected signal in the rising period and the falling period, we still can calculate the real delay time or *OPD* between the two interfering waves, as well as the speed of a moving object. However, in practice, accurately measuring the average frequency of the sinusoidal-wave beat signal is difficult. In particular, at the moments corresponding to the peaks or valleys of the modulation waveform, the frequency of the beat signal approaches zero and the measurement accuracy usually is the worst. This is the major weakness of sinusoidal-wave optical FMCW interference.

2.3 Multiple-Beam Optical FMCW Interference

Thus far, we have discussed a number of situations in which two optical FMCW waves are combined to interfere (i.e., *double-beam optical FMCW interference*). There are, however, some circumstances under which more than two optical FMCW waves are involved in the interference (i.e., *multiple-beam optical FMCW interference*).

If a number of optical FMCW waves, derived from the same light source and propagated along different paths, are recombined to interfere, the intensity of the resulting electric field can be expressed by

$$I(\tau_1,...\tau_m,t) = \left| \sum_{i=1}^{m} E_i(\tau_i,t) \right|^2$$

$$= \left| \sum_{i=1}^{m} E_{0i}e^{j[\phi_i(t-\tau_i)]} \right|^2,$$

(2-102)

where m is the total number of interfering waves, and τ_i, E_{0i}, and $\phi(t-\tau_i)$ are the propagation time, amplitude, and phase of the ith wave, respectively.

For instance, if three optical FMCW waves, derived from the same light source and propagated along different paths, are recombined to interfere, the intensity of the resulting electric field can be written as

$$I(\tau_1,\tau_2,\tau_3,t)=\left|E_1(\tau_1,t)+E_2(\tau_2,t)+E_3(\tau_3,t)\right|^2$$

$$=[E_1(\tau_1,t)+E_2(\tau_2,t)+E_3(\tau_3,t)][E_1(\tau_1,t)+E_2(\tau_2,t)+E_3(\tau_3,t)]*$$

$$=E_1(\tau_1,t)E_1*(\tau_1,t)+E_2(\tau_2,t)E_2*(\tau_2,t)+E_3(\tau_3,t)E_3*(\tau_3,t)$$

$$+E_1(\tau_1,t)E_2*(\tau_2,t)+E_1*(\tau_1,t)E_2(\tau_2,t)+E_2(\tau_2,t)E_3*(\tau_3,t)$$

$$+E_2*(\tau_2,t)E_3(\tau_3,t)+E_1(\tau_1,t)E_3*(\tau_3,t)+E_1*(\tau_1,t)E_3(\tau_3,t)$$

$$=I_1+I_2+I_3+2\sqrt{I_1I_2}\,\cos[\phi_1(t-\tau_1)-\phi_2(t-\tau_2)]$$

$$+2\sqrt{I_2I_3}\,\cos[\phi_2(t-\tau_2)-\phi_3(t-\tau_3)]$$

$$+2\sqrt{I_1I_3}\,\cos[\phi_1(t-\tau_1)-\phi_3(t-\tau_3)]$$

$$\text{(2-103)}$$

where I_1, I_2, and I_3 are the intensities of the three waves, respectively ($I_1 = E_1(\tau_1, t)E_1*(\tau_1, t) = E_{01}{}^2$, $I_2 = E_2(\tau_2, t)E_2*(\tau_2, t) = E_{02}{}^2$, $I_3 = E_3(\tau_3, t)E_3*(\tau_3, t) = E_{03}{}^2$); τ_1, τ_2, and τ_3 are the propagation times of the three waves, respectively.

If the angular frequency of the light source is modulated with a sawtooth waveform, the intensity of the resulting electric field in each modulation period can be written as

$$I(\tau_1,\tau_2,\tau_3,t)=I_1+I_2+I_3+2\sqrt{I_1I_2}\,\cos[\alpha(\tau_2-\tau_1)t-\omega_0(\tau_2-\tau_1)]$$

$$+2\sqrt{I_2I_3}\,\cos[\alpha(\tau_3-\tau_2)t-\omega_0(\tau_3-\tau_2)]$$

$$+2\sqrt{I_1I_3}\,\cos[\alpha(\tau_3-\tau_1)t-\omega_0(\tau_3-\tau_1)]$$

$$\text{(2-104)}$$

Apparently, the beat signal produced by three sawtooth-wave FMCW waves contains three beat components: the first one is the beat signal produced by the first and second waves ($2\sqrt{I_1I_2}\,\cos[\alpha(\tau_2-\tau_1)t-\omega_0(\tau_2-\tau_1)]$), the second one is the beat signal produced by the second and third waves ($2\sqrt{I_2I_3}\,\cos[\alpha(\tau_3-\tau_2)t-\omega_0(\tau_3-\tau_2)]$), and the last one is produced by the first and third waves ($2\sqrt{I_1I_3}\,\cos[\alpha(\tau_3-\tau_1)t-\omega_0(\tau_3-\tau_1)]$). When changing the propagation time of any wave, two beat components will be influenced.

Similarly, if n sawtooth-wave optical FMCW waves interfere, there will be $n(n-1)/2$ beat components in the detected signal. Altering the propagation time of any wave, $(n-1)$ beat components in the beat signal will be influenced. Hence, because of the mutual interference among the waves and the complexity of the beat signal, multiple-beam optical FMCW interference is rarely used in practice.

2.4 Multiple-Wavelength Optical FMCW Interference

We know that optical waves with different wavelengths (i.e., different distinct frequencies) from different light sources are incoherent. If a double-beam FMCW interferometer is illuminated by a number of light sources with different wavelengths, the intensity of the resulting electric field will be equal to the sum of the beat signals produced by all individual light sources,

$$I(\lambda_1,...\lambda_m,\tau,t) = \sum_{i=1}^{m} I_{0\lambda_i}\{1 + V_{\lambda_i}\cos[\phi_{\lambda_i}(t) - \phi_{\lambda_i}(t-\tau)]\} \qquad (2\text{-}105)$$

where m is the number of the light sources, $I_{0\lambda i}$, $V_{\lambda l}$, and $[\phi_{\lambda i}(t) - \phi_{\lambda i}(t-\tau)]$ are the average intensity, contrast, and phase of the beat signal produced by the ith light source, respectively, and τ is the delay time of the signal wave with respect to the reference wave in the double-beam FMCW interferometer.

In the case of double-wavelength illumination, for instance, the intensity of the resulting electric field will be

$$I(\lambda_1,\lambda_2,\tau,t) = I_{0\lambda_1}\{1 + V_{\lambda_1}\cos[\phi_{\lambda_1}(t) - \phi_{\lambda_1}(t-\tau)]\}$$
$$+ I_{0\lambda_2}\{1 + V_{\lambda_2}\cos[\phi_{\lambda_2}(t) - \phi_{\lambda_2}(t-\tau)]\} \qquad (2\text{-}106)$$

where $I_{0\lambda 1}$, $V_{\lambda 1}$ and $[\phi_{\lambda 1}(t) - \phi_{\lambda 1}(t-\tau)]$ are the average intensity, contrast, and phase of the beat signal produced by the first light source, respectively; and $I_{0\lambda 2}$, $V_{\lambda 2}$, and $[\phi_{\lambda 2}(t) - \phi_{\lambda 2}(t-\tau)]$ are the average intensity, contrast, and phase of the beat signal produced by the second light source, respectively.

If the optical angular frequencies of the two light sources are modulated with a sawtooth waveform of the same frequency but at different central angular frequencies and with different angular frequency modulation rates, the intensity of the resulting electric field in each modulation period becomes

$$I(\tau,t) = I_{01}[1 + V_1\cos(\alpha_1\tau t + \omega_{01}\tau)] + I_{02}[1 + V_2\cos(\alpha_2\tau t + \omega_{02}\tau)] \qquad (2\text{-}107)$$

where I_{01}, V_1, α_1 and ω_{01} are the average intensity, contrast, angular frequency modulation rate, and central angular frequency related to the first light source, respectively; and I_{02}, V_2, α_2, and ω_{02} are the average intensity, contrast, angular frequency modulation rate, and central angular frequency related to the second light source, respectively.

For simplicity, assuming $I_{01} = I_{02} = I_0/2$ and $V_1 = V_2 = V$, Equation (2-107) can be simplified as

$$I(\tau,t) = I_0 \left\{ 1 + \frac{V}{2}[\cos(\alpha_1 \tau t + \omega_{01}\tau) + \cos(\alpha_2 \tau t + \omega_{02}\tau)] \right\}$$

$$= I_0 \left\{ 1 + V \left[\cos\left(\frac{\alpha_1 - \alpha_2}{2}\tau t + \frac{\omega_{01} - \omega_{02}}{2}\tau \right) \cos\left(\frac{\alpha_1 + \alpha_2}{2}\tau t + \frac{\omega_{01} + \omega_{02}}{2}\tau \right) \right] \right\}$$

$$= I_0 \left\{ 1 + V \left[\cos\left(\frac{\Delta\alpha}{2}\tau t + \frac{\Delta\omega_0}{2}\tau \right) \cos\left(\overline{\alpha}\tau t + \overline{\omega}_0\tau \right) \right] \right\}$$

$$\text{(2-108)}$$

where $\Delta\alpha$ is the difference in the angular frequency modulation rates ($\Delta\alpha = \alpha_1 - \alpha_2$), $\Delta\omega_0$ is the difference in the central angular frequencies ($\Delta\omega_0 = \omega_{01} - \omega_{02}$), $\overline{\alpha}$ is the average value of the angular frequency modulation rates ($\overline{\alpha} = (\alpha_1 + \alpha_2)/2$), and $\overline{\omega}_0$ is the average value of the central angular frequencies ($\overline{\omega}_0 = (\omega_{01} + \omega_{02})/2$).

Equation (2-108) shows that the beat signal of the double-wavelength sawtooth-wave FMCW interference is characterized by the average angular frequency modulation rate and the average central angular frequency, but its amplitude is modulated by an envelope wave (usually called a *synthetic wave*), as shown in Figure 2.17. The period of the synthetic wave T_s equals

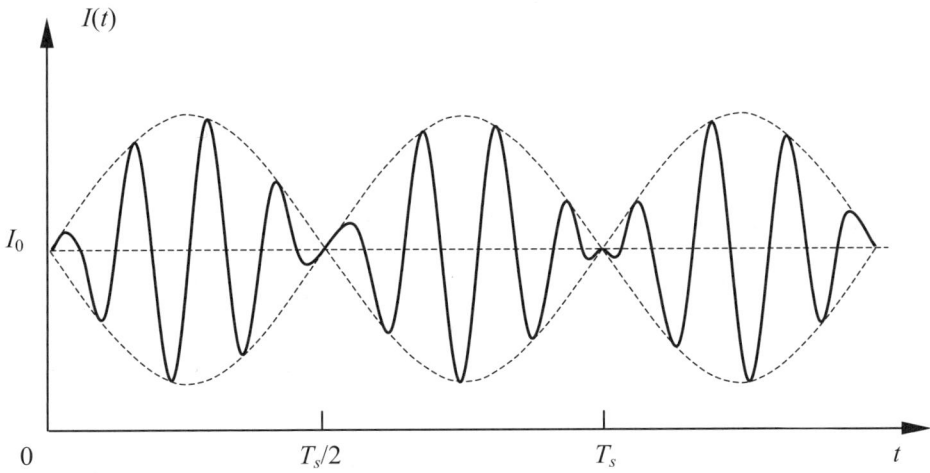

Figure 2.17. Waveform of the beat signal from a double-wavelength sawtooth-wave FMCW interferometer.

$$T_s = \frac{4\pi}{\Delta\alpha\tau}$$

$$= \frac{4\pi c}{\Delta\alpha OPD} ,$$

(2-109)

where c is the speed of light in free space; the initial phase of the synthetic wave ϕ_{s0} equals

$$\phi_{s0} = \frac{\Delta\omega_0\tau}{2}$$

$$= \frac{\Delta\omega_0 OPD}{2c} .$$

(2-110)

When the delay time or *OPD* in the double-wavelength FMCW interferometer changes, the waveforms of both the beat signal and the synthetic wave will shift along the temporal axis. The change of the optical path difference, which makes the synthetic wave shift by half a period (i.e., the signal contrast varies by one cycle), usually is defined as the wavelength of the synthetic wave. From Equation (2-110), it can be derived that the synthetic wavelength λ_s equals

$$\lambda_s = \frac{\lambda_{01}\lambda_{02}}{|\lambda_{01} - \lambda_{02}|},$$

(2-111)

where λ_{01} is the central wavelength of the first light source ($\lambda_{01} = 2\pi c/\omega_{01}$), and λ_{02} is the central wavelength of the second light source ($\lambda_{02} = 2\pi c/\omega_{02}$).

The synthetic wavelength generally is much longer than the optical wavelengths. This feature of the double-wavelength FMCW interferometer may be useful in practice. For instance, the synthetic wavelength can be used to measure the variation of the step-changed optical path difference, such as the step height, which is longer than the optical wavelength. In this case, the single-wavelength FMCW interferometer is not effective because the number of the full periods of the phase shift has been lost.

Note that the frequency of the synthetic wave is proportional to both $\Delta\alpha$ and *OPD*. To observe at least half a period of synthetic wave in a modulation period, the *OPD* must be

$$OPD \geq \frac{c\omega_m}{\Delta\alpha},$$

(2-112)

where ω_m is the angular frequency of the modulation signal.

Particularly if $\Delta\alpha=0$ (for instance, the interfering waves are from a single double-wavelength sawtooth-wave FMCW light source, such as a two-mode semiconductor laser), we will not be able to see the pattern of a synthetic wave, no matter how long the *OPD* is. Under this situation, the intensity of the resulting electric field becomes

$$I(\tau,t)=I_0\left\{1+V\left[\cos\left(\frac{\Delta\omega_0}{2}\tau\right)\cos(\alpha\tau t+\overline{\omega_0}\tau)\right]\right\}.$$

(2-113)

Obviously, the angular frequency of the beat signal is the same as that of the single-wavelength beat signal, but the initial phase of the beat signal is proportional to the average central angular frequency, and the contrast of the beat signal is modified by the factor $\cos(\Delta\omega_0\tau/2)$ or $\cos(\pi OPD/\lambda_s)$. Notice that the contrast of the beat signal is time-independent but periodically varies with the *OPD*. The change of the *OPD*, which makes the signal contrast vary by a period, is still equal to the synthetic wavelength λ_s, as shown in Figure 2.18.

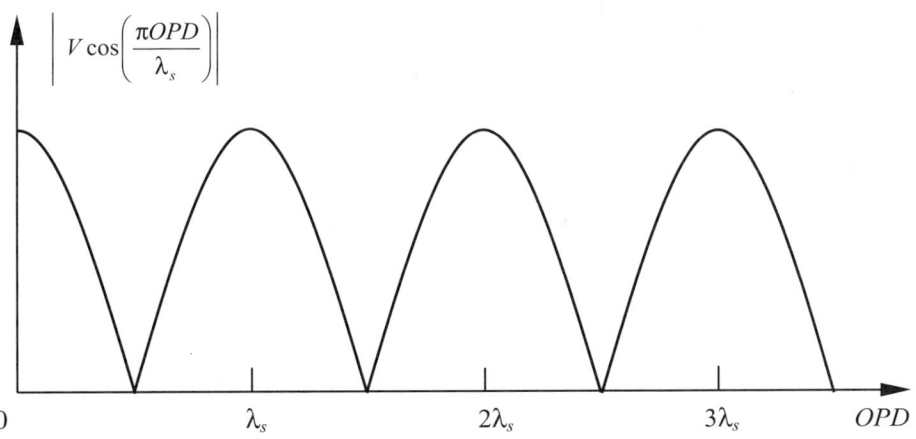

Figure 2.18. Contrast of the signal from a double-wavelength sawtooth-wave FMCW interferometer, when $\alpha_1 = \alpha_2$ and $\omega_{01} \neq \omega_{02}$.

Chapter 3

Optical Sources for Optical Frequency-Modulated Continuous-Wave Interference

Optical sources are very important to optical interference. They dominate the performance of optical interferometric systems. Optical FMCW interference requires coherent optical source whose frequency can be continuously modulated. In practice, only lasers are capable of satisfying this specific requirement.

In this chapter, we first briefly introduce some basic knowledge about optical sources, then discuss laser principles (including the concepts of stimulated emission, population inversion, optical resonators, laser modes, and frequency modulation), and finally discuss the most commonly used semiconductor lasers (including their driving circuit design, laser noise analysis, frequency drift, and feedback light effects).

3.1 Introduction to Optical Sources

Optical sources can be categorized into incandescent sources, luminescent sources, and lasers. *Incandescent sources*, such as tungsten-filament lamps, are based on the phenomenon of thermal radiation (i.e., objects emitting light by losing the kinetic energy of their constituent particles). Thermal radiation gives a continuously spread spectrum that can be described by using the parameter of temperature. Incandescent sources are the typical incoherent optical sources.

Luminescent sources, such as electric discharge lamps and light-emitting diodes, emit light by using the inner energy of the constituent particles and give discrete spectra or specific limited-band spectra that generally cannot be described by using the parameter of temperature. Some luminescent sources are frequently used as partially coherent optical sources.

Lasers are based on the phenomenon of stimulated emission and produce a highly directional, highly intense beam. The single-longitudinal-mode lasers have an extremely narrow spectral bandwidth and can be used as ideal coherent optical sources. Various types of lasers are now available and have been widely used in industrial processes, precision measurements, optical holography, optical communications, and medical treatments.

For optical FMCW interference, the optical source should meet the following essential requirements:

(1) Changeable frequency: The optical frequency of the optical source must be capable of being continuously modulated.

(2) High coherence: The coherence length of the optical source determines the measurement range of the interferometric system.

(3) Large frequency modulation excursion: A large optical frequency modulation excursion is essential to produce a high-frequency beat signal, leading to a high resolution of the optical path difference.

(4) High modulation rate: The modulation rate is also related to the beat frequency and dominates the dynamic characteristics of the system.

(5) Linear frequency modulation: Linear FMCW interference generally provides higher accuracy than nonlinear FMCW interference.

(6) Low noise: The deviation of the real optical wave from the ideal waveform should be as small as possible.

In addition, other factors, such as the lasing efficiency, physical size, reliability, durability, and cost, are also important to the real application.

In practice, only lasers are capable of satisfying the requirements above, and thus far semiconductor lasers are believed to be the best optical sources for optical FMCW interference because they have a number of superior features, such as high efficiency, direct current modulation, acceptable modulation linearity, high modulation rate, large modulation excursion, compact size, and low cost. In the following two sections, we will first introduce the general principles of lasers and then apply these concepts to semiconductor lasers.

3.2 Laser Principles

Laser is an acronym for light amplification by stimulated emission of radiation. To understand the operating principles of lasers, we have to understand what is meant by stimulated emission and under what conditions we can achieve the amplification of light by stimulated emission.

3.2.1 Energy Levels in Substances

We know that all substances are made up of molecules or atoms. In the energy domain, these particles can be characterized by discrete *energy levels* (or *states*). The lowest level has the lowest energy and usually is referred to as the *ground state*, while the upper levels have a higher energy and usually are referred to as the *excited states*. Each constituent molecule or atom can exist in one of the allowed levels, but normally it prefers to stay in the lower levels.

For a nondegenerate substance in thermal equilibrium, the population density N (i.e., the number of particles per unit volume) in an energy level E at a temperature T is given by the Maxwell-Boltzmann statistics

$$N = N_0 e^{-E/k_B T},$$

(3-1)

where N_0 is a constant for a given temperature, and k_B is the Boltzmann constant ($k_B = 1.38 \times 10^{-23}$ JK^{-1}). If we consider only two energy levels (E_i and E_j), the ratio of the population densities of the two levels would be

$$\frac{N_j}{N_i} = \frac{e^{-E_j/k_B T}}{e^{-E_i/k_B T}}.$$

(3-2)

In other words, the relative population density of the E_j relative to the E_i is

$$N_j = N_i e^{-(E_j - E_i)/k_B T}.$$

(3-3)

Obviously, in thermal equilibrium, the lower energy level is more densely populated than the higher energy level, and the population density has an exponential dependence on energy.

3.2.2 Optical Absorption and Emissions

Molecules or atoms can make upward or downward transitions between any two allowed states by absorbing or releasing an amount of energy, respectively. Such an amount of energy equals the difference between the two energy levels and can be in the form of light energy (i.e., photons), kinetic energy, or the inner energy of the particles.

Transitions involving emission or absorption of photons are called *radiative transitions*. If there are only two energy levels participating in an interaction with optical radiation, the photon frequency ν_{21} and the energy difference between the two energy levels ($E_2 - E_1$) are related by

$$h\nu_{21} = E_2 - E_1,$$

(3-4)

where h is the Planck constant ($h = 6.63 \times 10^{-34}$ J·s).

The radiative transitions between any two energy levels can be classified as the following three types.

(1) *Stimulated absorption*: A particle in a lower energy level (say level 1) can absorb a photon of frequency $(E_2-E_1)/h$ and make an upward transition to a higher energy level (level 2), as shown in Figure 3.1(a). The rate of stimulated absorption depends on both the number of particles present in the lower energy level and the energy density of radiation present in the substance.

(2) *Spontaneous emission*: A particle in a higher energy level (say level 2) can make a downward transition to a lower energy level (level 1) spontaneously by emitting a photon of frequency $(E_2-E_1)/h$, as shown in Figure 3.1(b). The rate of spontaneous transition depends only on the number of particles present in the higher energy level. The average value of the time period for which a particle can stay in a higher energy level before downward transition by spontaneous emission is called *lifetime*. The lifetime of an excited state may be very short (less then 10 ps) or may be fairly long (greater than 1 μs) depending on the property of the atomic system. An excited state with a long lifetime is referred to as a *metastable level*.

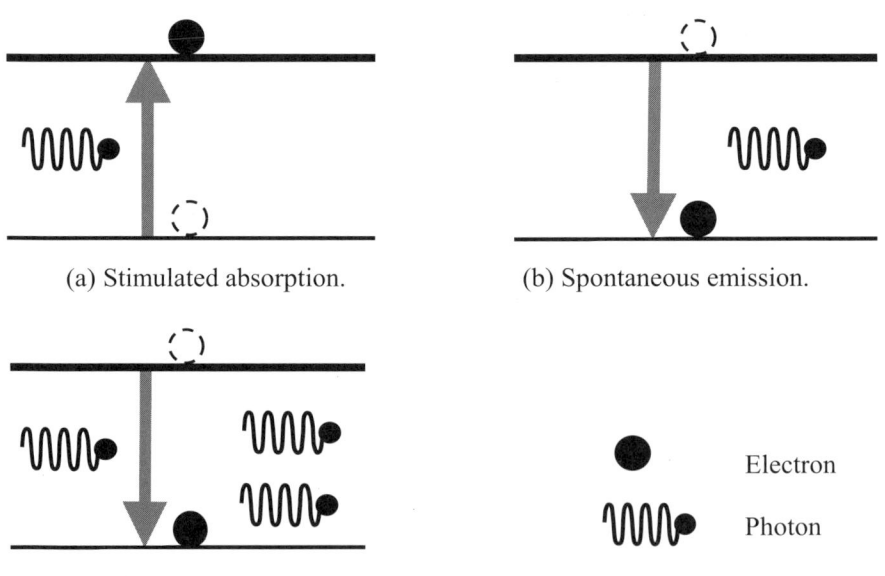

(a) Stimulated absorption.   (b) Spontaneous emission.

(c) Stimulated emission.

●   Electron

〜〜〜〜   Photon

Figure 3.1. Radiative transitions.

(3) *Stimulated emission*: A particle in a higher energy level (say level 2) can also make a downward transition in the presence of an external radiation of frequency $(E_2 - E_1)/h$ by emitting a photon of the same frequency, as shown in Figure 3.1(c). The rate of stimulated emission depends on both the number of particles in the higher energy level and the energy density of the external radiation.

Notice that the most important aspect of stimulated emission is that the emitted photon is identical with the stimulating photon. They have the same frequency, same polarization state, and same propagation direction. Therefore, stimulated emission can produce an amplification of the stimulating radiation.

3.2.3 Active Medium and Population Inversion

The substance in which light amplification occurs generally is referred to as the *active medium*. The active medium may be a collection of molecules, atoms, or ions and may be in the solid, liquid, or gaseous form, which usually is addressed as an *atomic system*.

In an atomic system, in the presence of external radiation, the probability of absorption per atom is the same as the probability of stimulated emission per atom. For instance, in a two-level (the ground state and an excited state) atomic system, if $\rho(\nu_{21})$ represents the radiation energy density of the frequency ν_{21}, the rate of absorption dN_1/dt would be

$$\frac{dN_1}{dt} = -B_{12}\rho(\nu_{21})N_1 \quad , \tag{3-5}$$

where B_{12} is the Einstein stimulated absorption coefficient, and N_1 is the population density in the lower energy level (level 1). The rate of emission dN_2/dt would be

$$\frac{dN_2}{dt} = -[A_{21} + B_{21}\rho(\nu_{21})]N_2 \quad , \tag{3-6}$$

where A_{21} is the Einstein spontaneous emission coefficient ($A_{21} = 8\pi h\nu_{21}^3 B_{21}/c^3$, where c is the speed of light in free space), B_{21} is the Einstein stimulated emission coefficient ($B_{21} = B_{12}$), and N_2 is the population density in the higher energy level (level 2). At steady state, the rate of absorption must be equal to the rate of emission; that is,

$$B_{12}\rho(\nu_{21})N_1 = [A_{21} + B_{21}\rho(\nu_{21})]N_2 \quad . \tag{3-7}$$

If we ignore spontaneous emission for the moment because we are interested in the amplification process due to stimulated emission, we can see that, for the rate of emission to exceed the rate of absorption of photons, N_2 must be greater than N_1. However, such a population distribution contradicts the normal population distribution in the atomic system in thermal equilibrium. This usually is called *population inversion*. Hence, population inversion is the necessary condition for light amplification.

Considering $A_{21} = 8\pi h v_{21}^3 B_{21}/c^3$ and $B_{21} = B_{12}$, Equation (3-7) can be rearranged as

$$\frac{N_2}{N_1} = \frac{B_{12}\rho(v_{21})}{A_{21} + B_{21}\rho(v_{21})}$$

$$= \frac{1}{1 + \dfrac{8\pi h v_{21}^3}{c^3 \rho(v_{21})}} \tag{3-8}$$

Obviously, the value of the right-hand side of this equation is always smaller than one. In other words, for a two-level atomic system, increasing the radiation energy density by an external radiation cannot cause population inversion. In some other atomic systems and under certain conditions, however, the population distribution may be inverted between a pair of levels.

For instance, in the three-level (the ground state and two excited states) atomic system illustrated in Figure 3.2(a), when the system is irradiated with a radiation of frequency v_{31} ($v_{31} = (E_3-E_1)/h$), the atoms can get excited from level 1 to level 3. (Excitation of an atomic system by an external means is often called *pumping*. Pumping can be done optically by using a specific lamp, electrically by electric discharge, chemically by chemical reaction, etc.) The excited atoms in level 3 can make a downward transition to level 1 by spontaneous and stimulated emission. This transition may be achieved either by a one-step transition from level 3 to level 1 or by a two-step transition from level 3 to level 2 and then to level 1. In the latter case, at a given pumping rate, if the rate of transition from level 3 to level 2 exceeds the rate of transition from level 2 to level 1 (this requires level 2 to be in a metastable state), then atoms would start accumulating in level 2, and the population of level 2 would increase with the pumping rate. When the pumping rate exceeds a certain threshold, the population in level 2 will become higher than that in level 1 and, consequently, the radiation at frequency $v_{21} = (E_2-E_1)/h$ will get amplified by stimulated emission.

In the three-level system above, population inversion is achieved between an excited state (level 2) and the ground state (level 1). Because the population of the ground state in thermal equilibrium is much larger than that in any excited

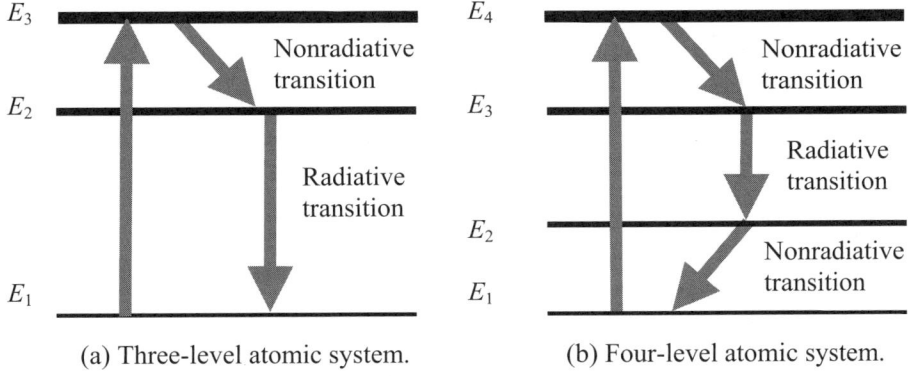

(a) Three-level atomic system.   (b) Four-level atomic system.

Figure 3.2. Energy level diagrams of active media.

state, the required threshold pumping power for population inversion generally is quite high.

Now, let's consider the four-level (the ground state and three excited states) atomic system illustrated in Figure 3.2(b). Assuming radiative transition takes place between levels 3 and 2, if the rate of transition from level 4 to level 3 and the rate of transition from level 2 to level 1 are large compared with the rate of radiative transition from level 3 to level 2, population inversion can be achieved between levels 3 and 2. Since the lower level of the radiative transition is no longer the ground state, the required threshold pumping power for population inversion usually is much lower. Therefore, it is easier to achieve population inversion with the four-level system than with the three-level system. This is why, in practice, most lasers employ four-level atomic systems.

The gain coefficient g for amplification of radiation in an active medium is defined by

$$g = \frac{dI}{Idx},$$

(3-9)

where dI/I is the relative increment of the light intensity and dx is a tiny distance passed by the light. Neglecting spontaneous emission, g would be

$$g = \frac{(N_2 - N_1)B_{21}h\nu_{21}dt}{dx}$$

$$= \frac{(N_2 - N_1)B_{21}h}{\lambda_{21}}$$

$$= C_{21}(N_2 - N_1),$$

(3-10)

where dx/dt is the speed of light in the medium, λ_{21} is the optical wavelength in the medium, and C_{21} is a constant given by

$$C_{21} = \frac{B_{21}h}{\lambda_{21}}$$

$$= \frac{A_{21}\lambda_{21}^2}{8\pi}.$$ (3-11)

Atomic physics indicates that the atoms in an energy level are not of the same energy but obey the population statistics

$$N = \frac{1}{2\pi\tau_n} \frac{N_n}{(E - E_n)^2 / \hbar^2 + 1/4\tau_n^2},$$ (3-12)

where N_n is the total number of the atoms in the E_n level, τ_n is the lifetime of this level $(\tau_n = 1/A_n)$, and $\hbar$ is a constant $(\hbar = h/2\pi)$. Therefore, the gain in radiation actually is frequency-dependent. The gain spectrum $g(\nu)$ for amplification of radiation in an active medium is

$$g(\nu) = \frac{1}{2\pi\tau_{21}} \frac{C_{21}(N_2 - N_1)}{4\pi^2(\nu - \nu_{21})^2 + (1/2\tau_{21})^2},$$ (3-13)

where τ_{21} is the lifetime of the spontaneous emission from E_2 to E_1. Notice that $g(\nu)d\nu$ represents the gain coefficient of the radiation between frequencies ν and $\nu+d\nu$.

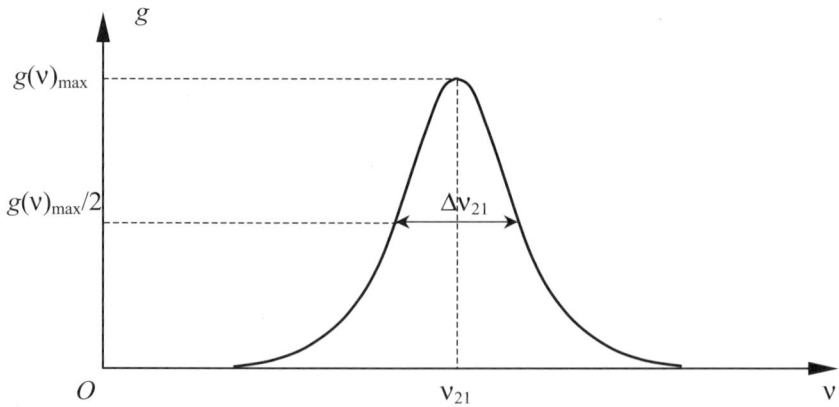

Figure 3.3. Lorentzian-linetype gain spectrum.

Figure 3.3 shows the curve of the gain spectrum $g(\nu)$. The curve is sharply peaked around a central resonant frequency ν_{21} and is usually named the Lorentzian linetype. The frequency interval between the two half-power points of the gain spectrum $g(\nu)$ is referred to as the natural linewidth, given by

$$\Delta\nu_{21} = \frac{1}{2\pi\tau_{21}}.$$

(3-14)

The gain spectrum may be broadened due to particle collision or movement (i.e., the Doppler effect). Collisional broadening is a type of homogeneous broadening, while Doppler broadening is a type of inhomogeneous broadening.

3.2.4 Optical Resonator and Laser Modes

From the previous subsection, we have learned that population inversion is necessary for optical amplification. However, to obtain a highly coherent, highly intense, and highly directional laser beam, the active medium has to be placed in an optical resonator.

The optical resonator (or optical resonant cavity), similar to a Fabry-Perot etalon, consists of two mirrors (M_1 and M_2) separated by a suitable distance, as shown in Figure 3.4. The mirrors may be planar or spherical and may either be attached to the ends of the active medium or be located outside it. Usually, one of the mirrors (say M_1) is almost totally reflecting and the other mirror (M_2) is partially reflecting so that a small fraction of optical energy can escape from the resonator as the output beam.

The optical resonator has two major functions. The first function is to provide a positive-feedback mechanism to the active medium. The light reflected back from the front and rear mirrors serves as positive feedback, so that an optical oscillation can build up and sustain.

Let R_1 and R_2 represent the reflectivities of the mirrors M_1 and M_2, respectively ($R_1 \approx 1$, $R_2 < 1$), and I_0 represent the intensity of the light beam leaving the mirror M_1 and propagating toward M_2. If α represents the average loss coefficient of the resonator due to all mechanisms other than the finite reflectivities of the mirrors, and g represents the gain coefficient of the resonator, then the optical intensity I_1 of the light beam after one complete round trip will be

$$I_1 = I_0 e^{-\alpha L} e^{gL} R_2 e^{-\alpha L} e^{gL} R_1$$
$$= I_0 R_1 R_2 e^{2(g-\alpha)L},$$

(3-15)

where L is the length of the resonant cavity. If the optical radiation has to build up and sustain in the resonator, we must have

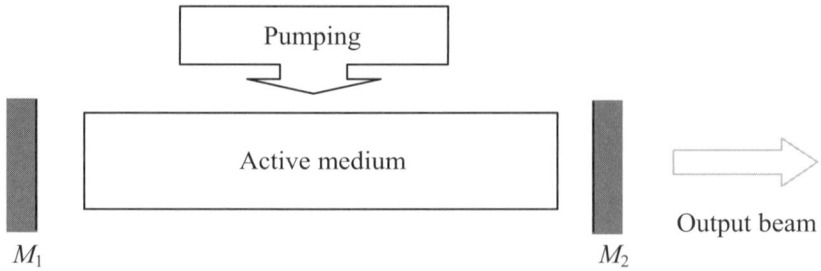

Figure 3.4. Structure of a laser.

$$R_1 R_2 e^{2(g-\alpha)L} \geq 1.$$
(3-16)

Generally, we define the *threshold condition* for laser oscillation as

$$R_1 R_2 e^{2(g-\alpha)L} = 1.$$
(3-17)

Correspondingly, the threshold gain coefficient g_{th} will be

$$g_{th} = \alpha - \frac{\ln(R_1 R_2)}{2L}.$$
(3-18)

When the gain coefficient becomes larger than the threshold value, the radiation will build up after propagating every round trip. However, this situation cannot last for long since increasing the power density induces more stimulated emissions, which results in faster depletion of the atoms from the excited state and reduces the population inversion and the gain coefficient. Soon, the radiation goes to a saturation state and the gain coefficient gets clamped at the threshold value.

The second function of the optical resonator is to establish the characteristic properties of the output beam. At the early stage, spontaneous and stimulated photons are emitted in all directions. Since most of them quickly escape from all sides of the active medium, only the axial beam continues to build up as it is reflected back and forth across the active medium. In addition, because the optical waves in the cavity are bouncing back and forth between the two mirrors, it is necessary that when a wave returns after one round trip it should be in phase with the existing wave. Therefore, it can be imagined that only a few frequencies in the gain spectrum can satisfy this resonant frequency requirement.

The optical resonator usually is characterized by transverse and longitudinal modes. *Transverse modes* refer to the specific transverse field distributions of the output beams, which depend on the physical geometry of the cavity and can be controlled by varying the mirror curvatures, cavity length, or cavity aperture. For instance, for a rectangular confocal cavity ($r_1 = r_2 = L$, where r_1 and r_2 are the curvature radii of the two mirrors and L is the cavity length), the electric field distribution in the stable state can be proved to be

$$E_{mn}(x, y) = C_{mn} H_m\left(\frac{\sqrt{2}x}{\omega}\right) H_n\left(\frac{\sqrt{2}y}{\omega}\right) e^{-\frac{x^2+y^2}{\omega^2}}, \qquad (3\text{-}19)$$

where C_{mn} is a constant, ω is a constant representing the cross-sectional radius of the laser beam, H_m and H_n are the m- and n-order Hermite polynomials, and m and n are positive integers ranging from zero to infinity that designate the different modes (generally named TEM$_{mn}$ modes).

Specifically, the intensity distribution $I(r)$ of the fundamental transverse (TEM$_{00}$) mode of a circular resonator can be approximately described by a Gaussian function,

$$I(r) = I_0 e^{-2r^2/\omega^2}, \qquad (3\text{-}20)$$

where I_0 is the optical intensity on the beam axis, and r is the radical coordinate, as shown in Figure 3.5.

The fundamental mode is characterized by a uniform phase front and has the least diffraction divergence, while the higher-order transverse modes have

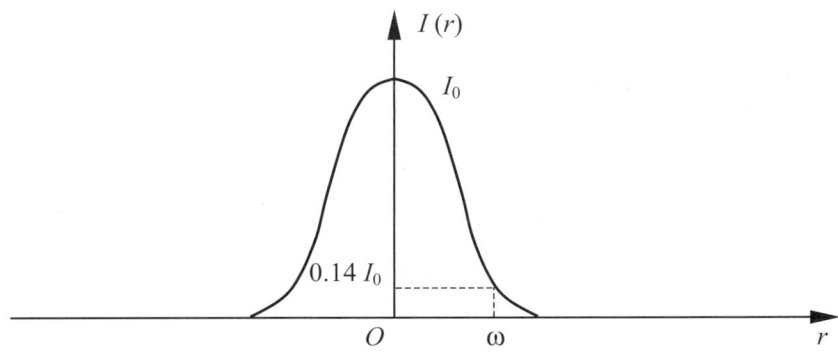

Figure 3.5. Gaussian intensity distribution of the fundamental transverse mode of a circular resonator.

phase reversals and zeroes across the beam. Hence, most lasers are designed to operate in the fundamental transverse mode.

Longitudinal modes refer to the specific wavelengths (or frequencies) of the output beam. Assuming that the active medium fills the entire cavity, if n represents the refractive index of the active medium and L represents the cavity length, the oscillation can take place only at the wavelength λ_k satisfying

$$\lambda_k = \frac{2nL}{k},$$

(3-21)

where k is an integer ($k = 1, 2, 3, \ldots$), or at the frequency v_k satisfying

$$v_k = \frac{kc}{2nL},$$

(3-22)

where c is the speed of light in free space. The gap between two consecutive resonant frequencies is

$$\Delta v_{k+1,k} = \frac{c}{2nL}.$$

(3-23)

It should be noted that the real lasing frequency is determined by both the gain spectrum of the active medium and the resonant frequencies of the optical cavity. In other words, only the frequencies satisfying both the threshold condition and the resonant condition can oscillate.

For an ordinary laser cavity without any specific restriction, there may be several frequencies within the gain spectrum that satisfy the resonant condition. As the pumping is increased, the longitudinal mode closest to the gain peak will be the first to start oscillation. When the pumping is further increased, other adjacent modes may also begin to oscillate, leading to multiple longitudinal mode oscillation, as shown in Figure 3.6(a). For a short optical resonant cavity, however, there may be only one frequency within the gain spectrum that satisfies the oscillating conditions, leading to single longitudinal mode oscillation, as shown in Figure 3.6(b).

Note that, even operating in single longitudinal mode, the laser still has a limited frequency bandwidth. If there is no active medium in the optical resonant cavity, the frequency bandwidth of the Fabry-Perot cavity δv_k is determined by

$$\delta v_k = \frac{c(1-R)}{2\pi nL\sqrt{R}},$$

(3-24)

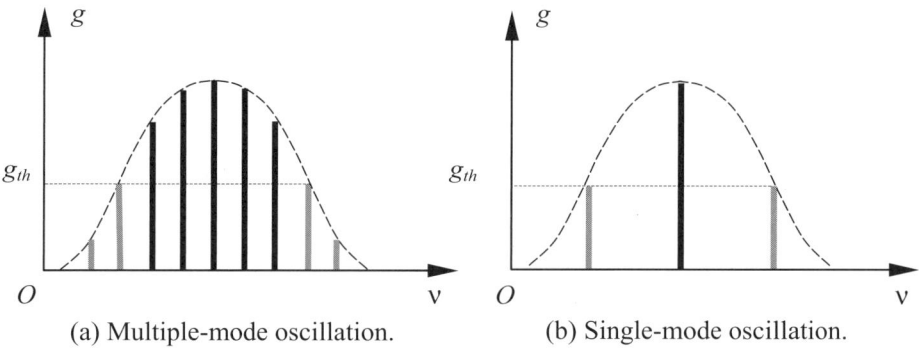

(a) Multiple-mode oscillation.   (b) Single-mode oscillation.

Figure 3.6. Longitudinal modes of lasers.

where c is the speed of light in free space, R is the reflectivity of the mirrors, n is the refractive index of the medium in the optical resonant cavity, and L is the length of the optical resonant cavity.

If there is an active medium in the cavity, the frequency bandwidth of the laser will get narrower. The limit of the laser bandwidth δv is given by

$$\delta v = 2\pi (\delta v_k)^2 \frac{h v_{21}}{P} \frac{N_2}{N_2 - N_1},$$

(3-25)

where P is the laser power, N_1 and N_2 are the population densities in energy levels 1 and 2, and v_{21} is the central frequency.

3.2.5 Frequency Modulation

There can be two approaches to modulating laser frequency: internal modulation and external modulation. *Internal modulation* is based on modulation of the properties of the optical resonator, while *external modulation* is based on modulation of the properties of the output laser beam.

Internal modulation can be realized by modulating the length of the optical resonant cavity, the refractive index of the medium in the cavity, or both. From the preceding subsection, we have seen that the length of the optical resonant cavity must be an integral number of optical wavelengths. When the cavity length is changed by a slight amount, the laser frequency will change to maintain an integral number. If L represents the cavity length and ΔL represents the change in cavity length, the change in optical wavelength $\Delta \lambda_k$ can be determined by

$$\Delta\lambda_k = \frac{\lambda_k \Delta L}{L},$$

(3-26)

where λ_k is the optical wavelength of the laser; or the change in frequency Δv_k can be determined by

$$\Delta v_k = \frac{-v_k \Delta L}{L},$$

(3-27)

where v_k is the optical frequency of the laser.

For instance, for a cavity of 1 meter, a change in mirror position of one wavelength will produce about 300 MHz in frequency shift. In practice, we often attach one of the mirrors with an acoustic transducer to modulate the cavity length and thus the laser frequency. The limitation of this method is that the modulation bandwidth is restrained by the mass of the transducer and mirror.

If we modulate the refraction index n of the active medium in the cavity (assuming the active medium fills the entire cavity) rather than the cavity length, the change in optical wavelength $\Delta\lambda_k$ would be

$$\Delta\lambda_k = \frac{\lambda_k \Delta n}{n},$$

(3-28)

where Δn is the change of the refraction index of the active medium, or the change in optical frequency Δv_k would be

$$\Delta v_k = \frac{-v_k \Delta n}{n}.$$

(3-29)

The modulation bandwidth of the index-modulation method is limited by the Q factor of the laser cavity ($Q = 2\pi v_{21}$(energy stored in the cavity)/(lost energy per unit time)). Since a light beam undergoes several reflections across the cavity, depending on the Q factor, before an appreciable portion of it is coupled out, the laser frequency must remain essentially constant during the transit time required for these multiple reflections. This limits the upper modulation frequency.

External modulation can be realized by using a Bragg frequency modulator or a phase modulator. The Bragg frequency modulator is based on the phenomenon of Bragg acousto-optic diffraction, which is believed to be the collision between photons and phonons. When a sinusoidal acoustic wave travels through some materials, it produces a stress wave, which, in turn, modulates the refractive index of the material and creates a traveling refractive index grating. If a laser beam projects on such a thick acoustic grating at the Bragg angle θ_B with respect to the wave fronts of the acoustic plane wave,

$$\sin \theta_B = \frac{\lambda}{2\Lambda} \,, \tag{3-30}$$

where λ is the optical wavelength in the medium, and Λ is the acoustic wavelength in the medium, the beam can be mostly coupled into the +1 order, and the optical frequency will be shifted by the acoustic frequency, as shown in Figure 3.7.

The drawback of the Bragg frequency modulator is that the Bragg angle is related to the acoustic wavelength. If the modulation bandwidth of the acoustic wave is broad, there will be a span in angle for which Bragg diffraction occurs. Moreover, the modulation bandwidth of the Bragg frequency modulator is limited by the acoustic transit time across the optical beam since the index grating in the optical beam at any moment in time must be of essentially constant frequency if all the light is to be diffracted at the same angle.

External frequency modulation can also be achieved by using a phase modulator because phase modulation and frequency modulation actually are the same in function—modulating angle. For instance, it is easy to prove that the phase modulation $[\phi(t) = \omega_0 t + A\sin(\omega_m t)]$ is equivalent to the frequency modulation $[\omega(t) = \omega_0 + A\omega_m\cos(\omega_m t)]$, where A is the phase modulation amplitude and $A\omega_m$ is the equivalent angular frequency modulation amplitude. The commonly used phase modulators include the mechanical phase modulator, fiber-optic phase modulator, and electro-optic phase modulator.

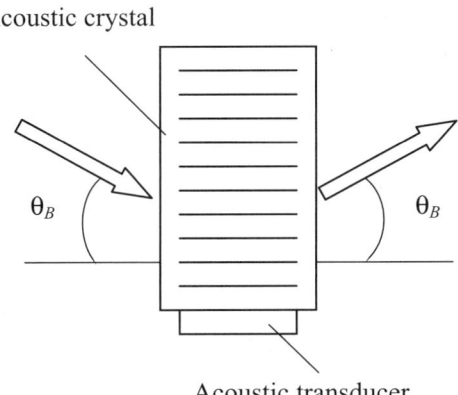

Figure 3.7. The Bragg frequency modulator.

The mechanical phase modulator can be a moving mirror (or a rotating glass plate). The light from the mirror is phase-modulated by the changes in the mirror position. This effect is often described in terms of the Doppler frequency shift, which is directly proportional to the mirror velocity and inversely proportional to the optical wavelength. Due to the inertia of the mirror mass, the mechanical phase modulator generally offers a low frequency-modulation bandwidth.

The fiber-optic phase modulator consists of a piezoelectric tube and a number of turns of single-mode optical fiber wrapped around the tube under a small tension, as shown in Figure 3.8. If a voltage is applied across the wall of a PZT tube, the circumference of the tube will change because of the strain induced in its wall thickness, height, and circumference by the electroconstriction effect, resulting in a change in the fiber length. The amplitude of the phase shift induced by the PZT tube fiber-optic phase modulator equals

$$\Delta\phi = k\Delta(n_e L)$$
$$= \eta NV \qquad ,$$

(3-31)

where k is the propagation number of the light beam in free space, n_e is the effective refractive index of the fiber, L is the total length of the fiber wrapped around the tube, N is the number of turns, V is the voltage applied to the PZT tube between the inner and outer surfaces, and η is the phase modulation efficiency.

The modulation amplitude can be largely increased by operating the modulator at one of the acoustic resonance frequencies of the PZT tube. These frequencies are determined by the dimensions of the tube, the material, and the

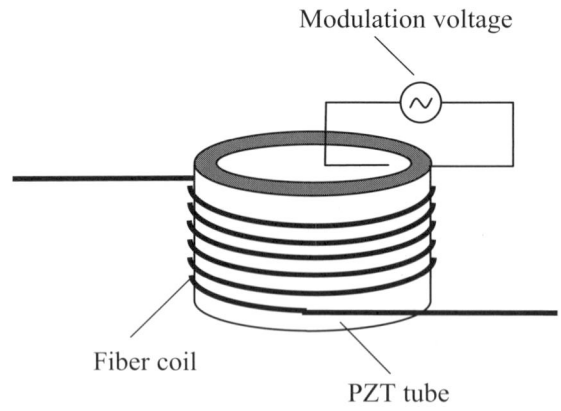

Figure 3.8. PZT tube fiber-optic phase modulator.

type of the acoustic resonant mode. The most commonly used acoustic resonance mode in a tube with a thin wall is the hoop mode, which corresponds to a
symmetrical expansion of the circumference. Other acoustic resonance modes
associated with the height and wall thickness of the tube can also be used since
expansion in these dimensions is also associated with a circumference change.
These acoustic resonance modes generally operate at high frequencies.

Due to their nonlinear frequency response, fiber-optic phase modulators are
normally used for sinusoidal-wave frequency modulation. Frequency modulation with other waveforms (such as sawtooth waves and triangular waves) is
not generally possible except at a low frequency (usually less than 1 kHz) because it requires a uniform frequency response for the significant harmonic frequency components of the waveform.

A single-mode fiber-optic phase modulator generally introduces a birefringence in the fiber due to tension coiling. This undesirable feature is particularly
destructive for some extremely sensitive phase measurements such as in fiber-
optic gyroscopes. In this circumstance, a polarization-maintaining fiber should
be used instead of the single-mode fiber.

The electro-optic phase modulator usually is based on the Pockels effect.
When an electric voltage is applied along the optical axis of an electro-optic
crystal, such as KDP (KH_2PO_4), KDA (KH_2AsO_4), or KD*P (KD_2PO_4), the refractive indexes of the two orthogonal principal axes will be altered.

Figure 3.9 shows a longitudinal electro-optic phase modulator that consists
of an input polarizer and an electro-optic crystal coated with a transparent electrode on both the front and back surfaces. The polarization direction of the polarizer is aligned parallel to one of the principal axes of the electro-optic crystal
(say the x-axis) so that the polarization direction of the laser beam stays the
same during the frequency modulation.

The Pockels effect is a linear electro-optic effect. The induced birefringence is proportional to the external electric field and therefore the applied
voltage. The phase shift of the modulator can be written as

$$\Delta\phi = k\Delta n_x L$$
$$= \frac{kn_x^3 r_{63} V}{2},$$

(3-32)

where k is the propagation number of the light in free space, n_x is the refractive
index in the x-axis without modulation, Δn_x is the variation of n_x with modulation, r_{63} is the electro-optic coefficient of the crystal, and V is applied voltage.

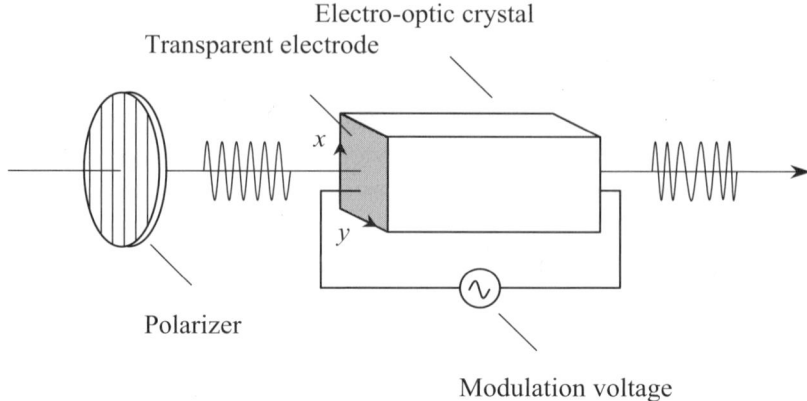

Figure 3.9. An electro-optic phase modulator.

3.3 Semiconductor Lasers (Laser Diodes)

Semiconductor lasers are laser devices that utilize semiconductor materials as the active medium for laser emission. They are the smallest lasers yet developed, typically about the size of a grain of salt. Since the basic structure of semiconductor lasers is a semiconductor *p-n* junction, semiconductor lasers are also called *laser diodes*.

3.3.1 Energy Bands in Semiconductors

Semiconductors, as the name indicates, are materials whose electrical conductivity values are between those of conductors and insulators. Most semiconductors are crystalline solids formed by some of the elements from group II to group VI in the periodic table of the elements. When N atoms of a semiconductor form a crystal, since the atoms are packed very closely, the electron clouds surrounding the positively charged nuclei overlap each other, allowing the valence electrons to move from one atom to another. From another point of view, the valence electrons now belong to the whole semiconductor rather than the individual atoms, and, as a result, each energy level of the atoms separates into N close energy levels. In practice, since the number of atoms in a semiconductor is extremely large, these close energy levels can be treated as a continuous energy band. Specifically, the valence energy level of an isolated atom becomes the *valence band* in the semiconductor crystal, and the other upper levels of an isolated atom become the *vacant bands* in the semiconductor crystal. The energy gaps between any two energy bands are named *forbidden bands*. In an intrinsic semiconductor at very low temperature ($0 \, ^\circ$K), the valence band is completely full and the vacant bands are completely empty of electrons. At higher temperatures, however, the thermal movement can bring some electrons from the valence band into the nearest vacant band. The vacant band with electrons is named a *conduction band*. In the valence band, the positions due to electrons leaving are called *holes*, as shown in Figure 3.10. Both electrons and holes are called *current carriers*.

Intrinsic semiconductors are perfect crystals (no defects or impurities) in which the energy gap between the conduction band and the valence band is in the range from a few tenths of an electron volt up to 2 eV. For an intrinsic semiconductor in thermal equilibrium, the occupational probability f_e of the allowed states of electrons is given by the Fermi distribution,

$$f_e(E) = \frac{1}{1 + e^{(E-E_F)/k_BT}},$$

(3-33)

where E_F is a constant called the *Fermi energy* (or *Fermi level*) that represents the energy value at which the probability of occupation of electrons is 0.5 (if there existed an allowed level at this energy value), and k_B is the Boltzmann constant. In an intrinsic semiconductor, the number of holes in the valence band is equal to the number of electrons in the conduction band, and the Fermi level lies approximately midway between the top of the valence band and the bottom of the conduction band, as shown in Figure 3.11(b).

Note that the Fermi level of the electrons can also be used to describe the distribution of the holes in the semiconductor, but the occupational probability of the holes f_h should be equal to

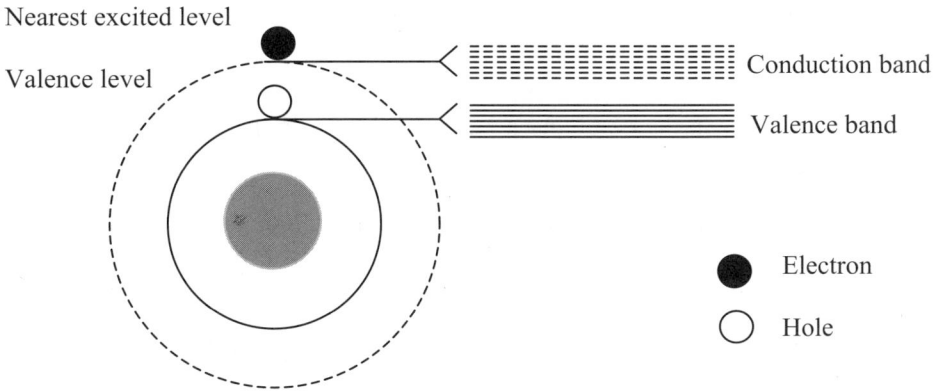

Figure 3.10. Energy bands in a semiconductor.

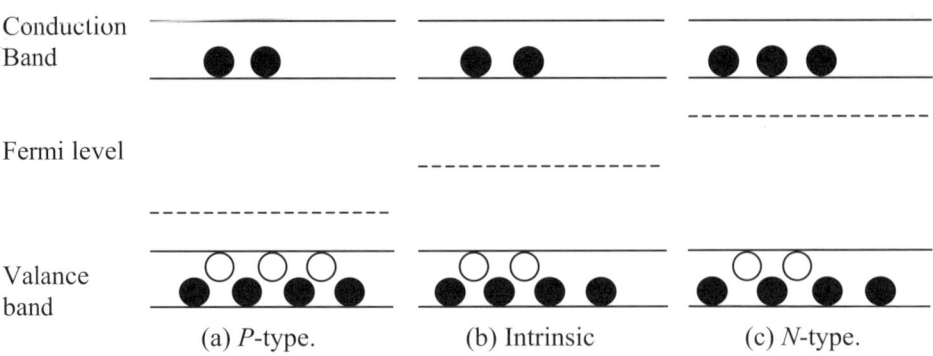

Figure 3.11. Distributions of electrons and holes in semiconductors.

$$f_h = 1 - f_e.$$
$$(3-34)$$

Extrinsic semiconductors are those materials in which a small amount of atoms in the crystal lattice are replaced by impurities. *Donor impurities* are atoms that have more valence electrons than required to complete the bonds with neighbor atoms, and the semiconductor containing donor impurities is said to be *n-type*. *Acceptor impurities* are atoms that have less valence electrons than required to complete the bonds with neighbor atoms, and the semiconductor containing acceptor impurities is said to be *p-type*. For extrinsic semiconductors in thermal equilibrium, since their electrical properties have been modified by impurities, the number of electrons in the conduction band and the number of holes in the valence band are not equal, and the Fermi levels are no longer located in the middle. Specifically, in a *p*-type semiconductor, there are a greater number of holes in the valence band than electrons in the conduction band, and therefore the Fermi level is located near the valence band, as shown in Figure 3.11(a); in an *n*-type semiconductor, however, there are a greater number of electrons in the conduction band than holes in the valence band, and therefore the Fermi level is located near the conduction band, as shown in Figure 3.11(c).

In a crystalline semiconductor, the atoms are arranged in a regular periodical lattice. The motion of an electron in such a crystal in any particular direction can be viewed as the motion of a negatively charged particle in a periodically varying electrostatic field produced by the atomic nuclei, and it can be described by using either the parameter momentum or the parameter propagation number k of the corresponding de Broglie wave. For instance, we usually use the dependence of the electron energy on the propagation number (*E-k diagram*) to describe the transition property of a semiconductor material. In particular, the extremes of the *E-k* curves generally are compatible with the energy band edges. For instance, the minimum value of the *E-k* curve of the conduction band corresponds to the bottom of the conduction band, while the maximum value of the *E-k* curve of the valence band corresponds to the top of the valence band.

Semiconductors in which the minimum value $(E_c)_{min}$ of the conduction band and the maximum value $(E_v)_{max}$ of the valence band occur at the same *k* value are known as *direct band-gap semiconductors*, as shown in Figure 3.12(a); semiconductors in which the minimum value $(E_c)_{min}$ of the conduction band and the maximum value $(E_v)_{max}$ of valence band occur at two different values of *k* are known as *indirect band-gap semiconductors*, as shown in Figure 3.12(b).

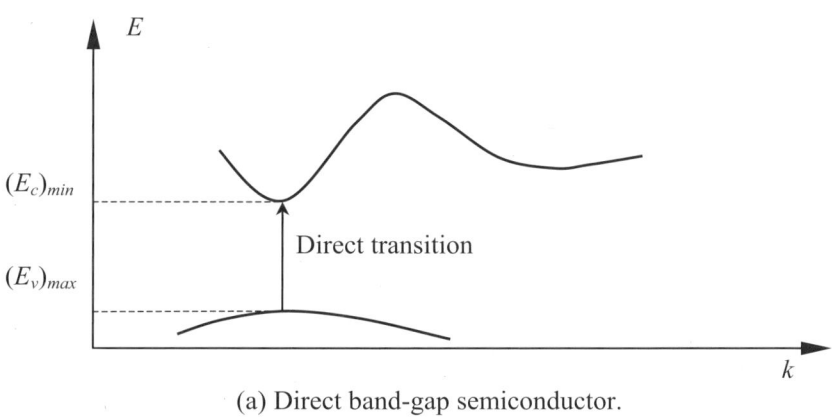

(a) Direct band-gap semiconductor.

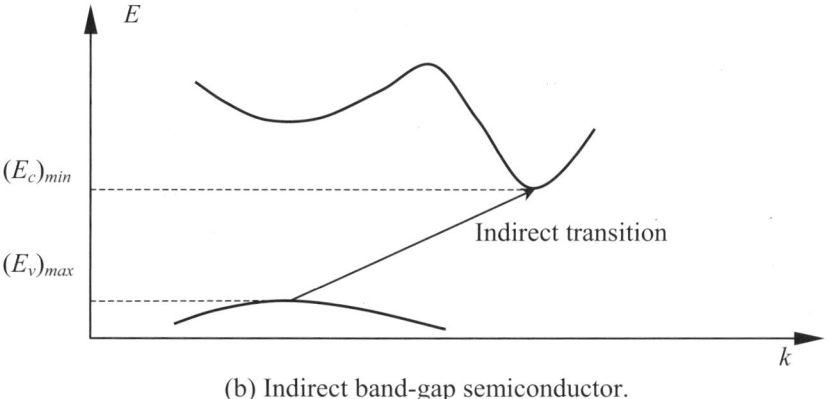

(b) Indirect band-gap semiconductor.

Figure 3.12. *E-k* diagrams of semiconductors.

3.3.2 Optical Absorption and Emissions in Semiconductors

The radiative transitions in a semiconductor generally take place between the valence band and the conduction band. Similar to the case of the isolated atomic system, there can be three different types of radiative transitions in a semiconductor. (1) An electron in the valence band can absorb a photon of frequency $(E_2 - E_1)/h$ and make an upward transition to the conduction band, where E_1 and E_2 are the energies associated with the initial and final states of the electron in the valence and conduction bands, respectively. (2) An electron in the conduction band can recombine with a hole in the valence band by spontaneously emitting a photon corresponding to the energy difference between its initial and final states. (3) An electron in the conduction band can also recombine with a hole in the valence band in the presence of an external radiation of frequency $(E_2 - E_1)/h$ by emitting a photon identical to the incident photon.

Different from the case of the isolated atomic system is the interaction of photons with electrons and holes in a semiconductor, which requires the satisfaction of both the energy conservation law and the momentum conservation law. The conservation of energy requires

$$E_1 + h\nu = E_2,$$
(3-35)

while the conservation of momentum requires

$$\hbar k_1 + \hbar k = \hbar k_2,$$
(3-36)

where k_1 and k_2 represent the propagation numbers associated with the electrons in the valence band and the conduction band, respectively, and k is the propagation number associated with the photon ($k = 2\pi/\lambda$). In practice, k is 2–3 orders smaller then k_1 and k_2, and therefore, the second condition can be simplified as

$$k_1 = k_2.$$
(3-37)

This equation implies that the allowed transitions between the conduction band and the valence band are vertical in the E–k diagram (called *direct transition*). However, this does not completely rule out the possibility of occurrence of the transitions that are not vertical in the E–k diagram (called *indirect transition*). Actually, if a third entity "phonon" participates in the interaction process, both the total energy and the total momentum can still be conserved. Phonons are quanta of lattice vibrations whose energy is relatively small but whose momentum can be quite large and comparable to that of electrons. Nevertheless, the probability of occurrence of the phonon-assisted transitions in a semiconductor is much smaller compared with the occurrence of the vertical transitions because the phonon-assisted transitions requires additional phonons in an appropriate quantity to offset the energy mismatch and the momentum mismatch. Therefore, almost all efficient semiconductor light sources are fabricated by using direct band-gap semiconductors, such as the binary compound semiconductors GaAs and InP. Silicon and germanium are indirect band-gap semiconductors, and thus they are not efficient radiation materials.

3.3.3 Active Medium and Population Inversion in Semiconductor Lasers

Now, let's consider a p-n junction formed between a p-type and an n-type GaAs semiconductor, as shown in Figure 3.13(a). Because the concentrations of electrons and holes in the p and n regions are different, electrons from the n region diffuse into the p region, and holes from the p region diffuse into the n region. The diffusion of these carriers across the junction leads to a built-in potential difference V_b between the positively charged ions on the n side and the negatively charged ions on the p side of the junction. This potential difference reduces the potential energy of the electrons on the n side with respect to that on the p side by

$$eV_b = (E_F)_n - (E_F)_p ,$$

(3-38)

where e is the magnitude of the electronic charge ($e = 1.6 \times 10^{-19}$ C), and $(E_F)_n$ and $(E_F)_p$ are the Fermi energies of the n-type and p-type semiconductors, respectively, as shown in Figure 3.13(b). The space-charge region is referred to as the *depletion region* because there is no free carrier there, and the width of the depletion region depends on the number of acceptor ions and donor ions on either side of the junction.

Note that the Fermi levels on both sides of the p-n junction are aligned at the same level. This is because in the absence of any applied external energy source, the charge neutrality in the material requires that the probability of finding an electron should be the same everywhere, hence, there should exist only one Fermi function in the semiconductor crystal.

If a forward bias voltage V is applied to the p-n junction, the potential energy of electrons on the n side of the junction increases, and correspondingly the energy bands move up. The Fermi levels on both sides of the junction separate by eV, as shown in Figure 3.13(c). The increased potential energy of the carriers pushes them into the depletion region, where they recombine to produce a forward current through the junction. The excess energy of the electrons may be released in the form of photons.

On each side of the *p-n* junction, because the incoming carriers are of the minority carriers on the side, the population of the minority carriers in the depletion region is significantly affected, and therefore the minority and majority carriers are no longer described by the same Fermi level. In other words, there are two Fermi levels (usually called *quasi-Fermi levels*) in the depletion region: one is for electrons and the other is for holes. For instance, on the *p* side, the Fermi level for holes basically remains the same because holes are the majority carriers and are affected less, while the Fermi level for electrons $(E_F)_{pe}$ moves up close to the Fermi level of the electrons $(E_F)_n$ in the *n*-region, as shown in Figure 3.13(c).

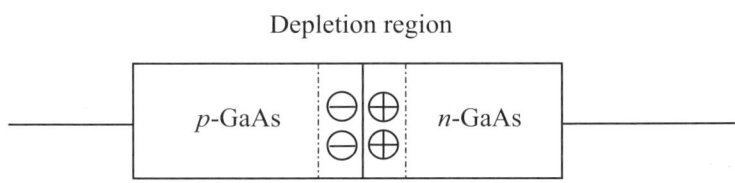

(a) Structure of the *p-n* junction.

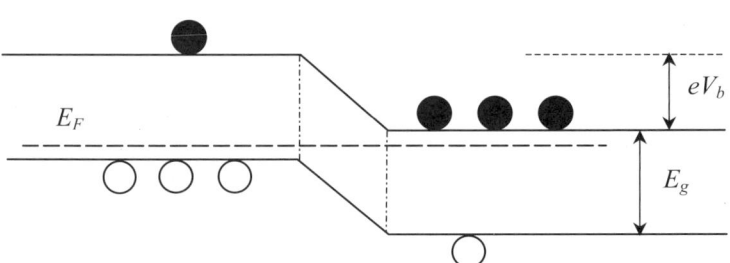

(b) Energy bands of an unbiased *p-n* junction.

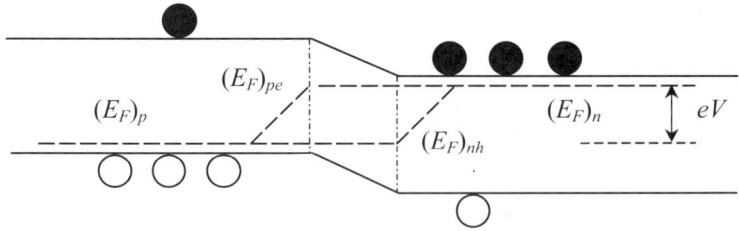

(c) Energy bands of a forward biased *p-n* junction.

Figure 3.13. Characteristics of a semiconductor *p-n* junction.

Note that, even though the separation between the quasi-Fermi levels is less than the band-gap energy E_g, there still can be light emission because of a forward current through the device. This is the basis of operation of the light-emitting diode (LED), and the device does not have a threshold value. However, for light amplification by stimulated emission, some special requirements have to be satisfied.

Let $\rho(v)$ represent the radiation energy density of the frequency v in the depletion region. The rate of increase of the photons due to stimulated emission dn_r/dt can be written as

$$\frac{dn_r}{dt} = B_{cv} n_{ce} n_{vh} \rho(v)$$

(3-39)

where B_{cv} is the Einstein stimulated emission coefficient, n_{ce} is the number of electrons in the conduction band, and n_{vh} is the number of holes in the valence band. The rate of decrease of the photons due to stimulated absorption dn_a/dt can be written as

$$\frac{dn_a}{dt} = B_{vc} n_{ve} n_{ch} \rho(v)$$

(3-40)

where B_{vc} is the Einstein stimulated absorption coefficient ($B_{vc} = B_{cv}$), n_{ve} is the number of electrons in the valence band, and n_{ch} is the number of holes in the conduction band. If we ignore spontaneous emission for the moment, we can find out that the rate of the net change of the photons dn/dt will be

$$\frac{dn}{dt} = \frac{dn_r}{dt} - \frac{dn_a}{dt}$$
$$= B_{cv} \rho(v)[n_{ce} n_{vh} - n_{ve} n_{ch}].$$

(3-41)

If the number of carriers is represented in terms of the energy density and the Fermi function of the electrons, we have

$$n_{ce} = N_c(E) f_{ce}(E),$$

(3-42)

$$n_{ch} = N_c(E)[1 - f_{ce}(E)],$$

(3-43)

$$n_{ve} = N_v(E - hv) f_{ve}(E - hv),$$

(3-44)

$$n_{vh} = N_v(E - h\nu)[1 - f_{ve}(E - h\nu)], \tag{3-45}$$

where $N_c(E)$ and $N_v(E)$ are the energy densities in the conduction band and the valence band, respectively, and $f_{ce}(E)$ and $f_{ve}(E)$ are the Fermi distributions of the electrons in the conduction band and the valence band, respectively. Substituting the equations above into Equation (3-41), we have

$$\frac{dn}{dt} = B_{cv}\rho(\nu)N_c(E)N_v(E - h\nu)[f_{ce}(E) - f_{ve}(E - h\nu)]. \tag{3-46}$$

For the rate of stimulated emission to exceed the rate of stimulated absorption of photons (i.e., $dn/dt > 0$), it is required that

$$f_{ce}(E) - f_{ve}(E - h\nu) > 0. \tag{3-47}$$

Considering the equations

$$f_{ce}(E) = \frac{1}{1 + e^{\frac{E - (E_F)_n}{k_B T}}}, \tag{3-48}$$

$$f_{ve}(E) = \frac{1}{1 + e^{\frac{E - (E_F)_p}{k_B T}}}. \tag{3-49}$$

Equation (3-47) can be simplified as

$$(E_F)_n - (E_F)_p \geq h\nu. \tag{3-50}$$

This equation implies that there is a large concentration of electrons in the conduction band and a large concentration of holes in the valence band, which is unlike the normal carrier distribution in a semiconductor in thermal equilibrium. For this reason, this is the requirement of carrier population inversion for a semiconductor laser, which is equivalent to the requirement of population inversion for an atomic system.

Since $h\nu \approx E_g$, the equation above can be rewritten as

$$(E_F)_n - (E_F)_p \geq E_g. \tag{3-51}$$

In general, it is not possible for a *p-n* junction formed by the moderately doped *p*-type and *n*-type semiconductors to satisfy this equation. However, if a *p-n* junction is formed by highly doped p^+-type and n^+-type semiconductors in which the Fermi levels are located inside the respective bands, and if a strong bias voltage ($V > E_g/e$) is applied across the junction, this equation can be satisfied, leading to light amplification by stimulated emission, as shown in Figure 3.14.

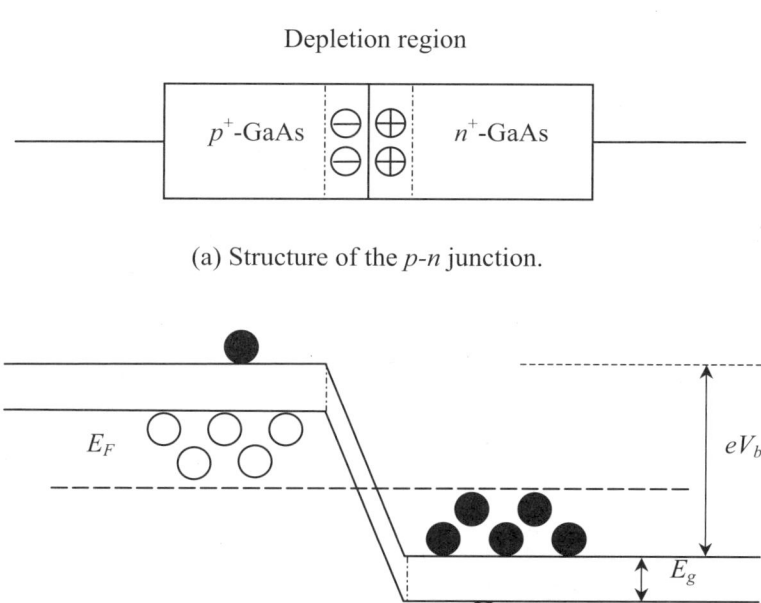

(a) Structure of the *p-n* junction.

(b) Energy bands of an unbiased *p-n* junction.

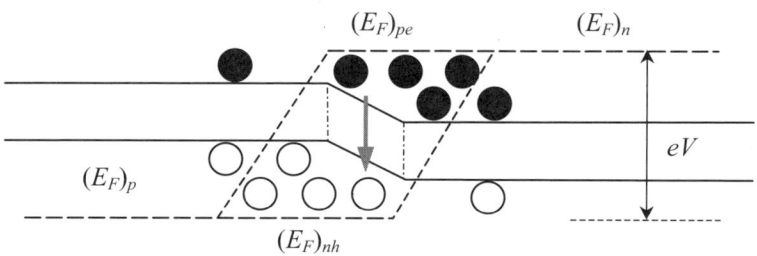

(c) Energy bands of a forward biased *p-n* junction.

Figure 3.14. Characteristics of a highly doped semiconductor *p-n* junction.

The first semiconductor laser was a *homojunction laser* and was invented as early as 1962. In this device, the *p-n* junction is formed by using the same semiconductor material on both sides of the junction. Homojunction semiconductor lasers can operate only in a pulse mode since the threshold current values are in the range of a few amperes to tens of amperes, which could lead to catastrophic damage to the devices if they operate continuously.

All modern semiconductor lasers use the *double heterojunction* structure, in which a thin layer (typically about 0.1 μm thick) of a suitable semiconductor is sandwiched between two layers of a higher-band-gap semiconductor, forming two heterojunctions. The narrower-band-gap thin layer, acting as the active region, generally has a higher refractive index. This structure provides so-called "carrier confinement" and "optical confinement", leading to low threshold current and high overall efficiency, as shown in Figure 3.15.

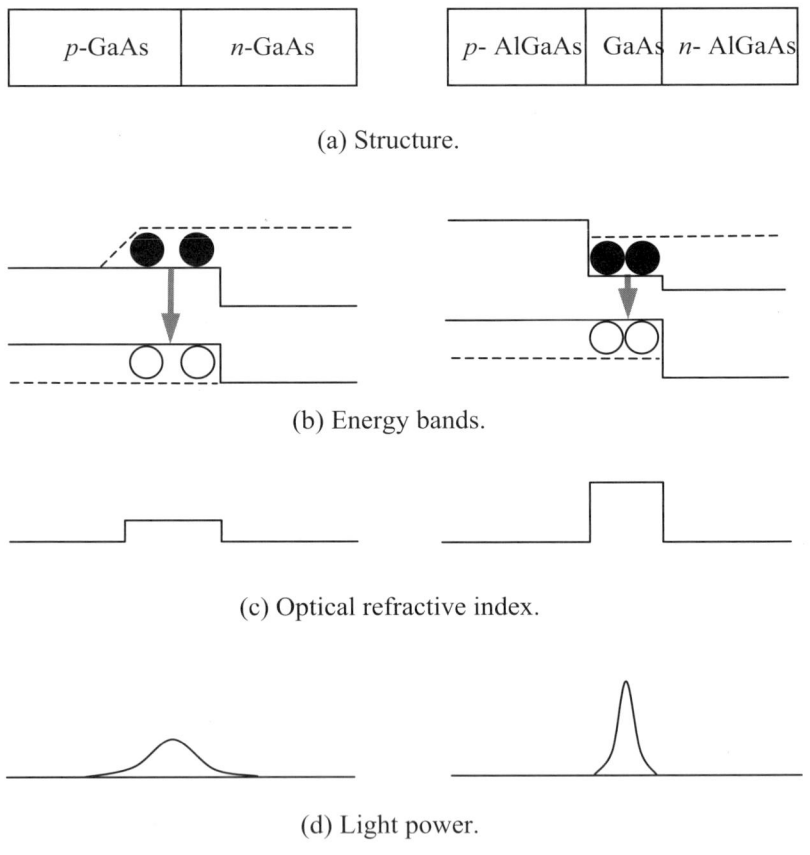

(a) Structure.

(b) Energy bands.

(c) Optical refractive index.

(d) Light power.

Figure 3.15. Comparison of homojunction structure and heterojunction structure (left—homojunction, right—double heterojunction).

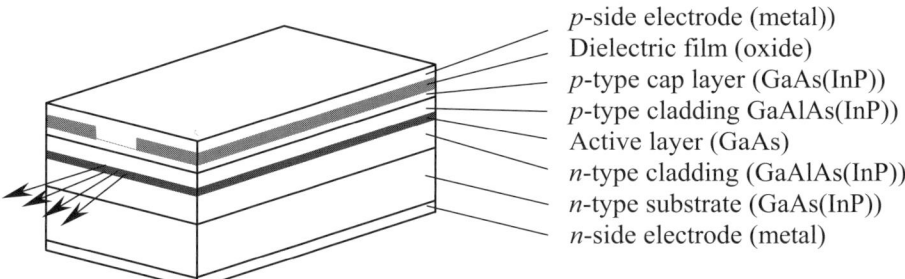

Figure 3.16. Typical structure of the modern semiconductor lasers.

Quantum-well (QW) lasers are special double-heterojunction semiconductor lasers that have a small active region (typically less than 200 Å). This small active region confines the excitation current to an even smaller lateral region so that the threshold current and also the heat can be further reduced.

Figure 3.16 shows a typical structure of modern semiconductor lasers. Many other structures for semiconductor lasers also exist. The spectral wavelength of semiconductor lasers depends on the property of the impurities in the semiconductor materials, typically InGaAlP devices lasing in the 630–690 nm wavelength region, GaAlAs devices lasing in the 780–870 nm region, InGaAs devices lasing in the 900–1020 nm region, and InGaAsP devices lasing in the 1.3–2.1 μm region.

Nearly all semiconductor lasers are pumped by the injection of electric current. The gain coefficient g of a semiconductor laser can be written as

$$g = \beta J ,$$

(3-52)

where β is a constant (called the gain factor) determined by the properties of the material and structure of the semiconductor laser and J is current density.

Figure 3.17 shows a typical variation of output power from a semiconductor laser as a function of the drive current. Below the threshold current, the output power is low. As the current crosses the threshold value, the output power increases significantly. The emission appearing below threshold is mainly due to spontaneous transitions, while above threshold it is primarily due to stimulated emission.

Semiconductor lasers are usually specified by the maximum current (or maximum output power P_{max}), which corresponds to 25–50% of the instant catastrophic failure current (or instant catastrophic failure output power) for long-term reliability. Operation at the catastrophic limit can cause permanent facet damage or junction degradation and thus destroy the laser in nanoseconds to microseconds.

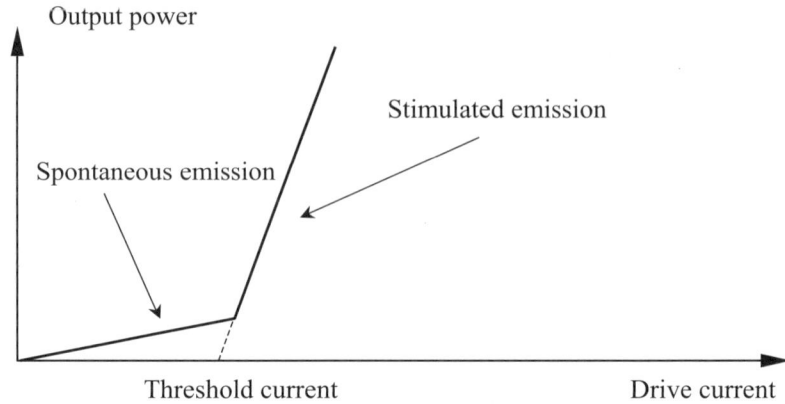

Figure 3.17. Typical output power-current relationship of the semiconductor laser.

3.3.4 Optical Resonator and Modes of Semiconductor Lasers

Unlike conventional lasers, semiconductor lasers normally use the two cleaved ends of the semiconductor crystal, rather than extra mirrors, to form an optical resonator. One of the cleaved ends may have a high-reflectivity coating, while another may have a partial reflectivity coating for laser output. The two other ends in the perpendicular direction are saw-cut to reduce reflections from these ends and prevent lasing along this direction or are surrounded by low-index materials to form a dielectric waveguide that can also be characterized by various transverse modes of propagation.

Most advanced semiconductor lasers use a strip waveguide cavity (typically 1–2 μm high, 3–7 μm wide, and 200–1000 μm long) and operate in the fundamental transverse mode (TEM$_{00}$), whose electric field distribution can be described by a Gaussian function with two different widths along the transverse direction w_t and along the lateral direction w_l,

$$E(x, y) = Ae^{-\left[\frac{x^2}{\omega_t^2} + \frac{y^2}{\omega_l^2}\right]}, \tag{3-53}$$

where A is a constant and x and y represent axes parallel and perpendicular to the p-n junction plane. The typical values of w_t and w_l are 0.5–1 μm and 1–2 μm, respectively. The far-field pattern of such output is elliptical, with a larger divergence in a plane perpendicular to the p-n junction, as shown in Figure 3.18. The divergences parallel and perpendicular to the junction plane are typically 5–10° and 30–50°, respectively.

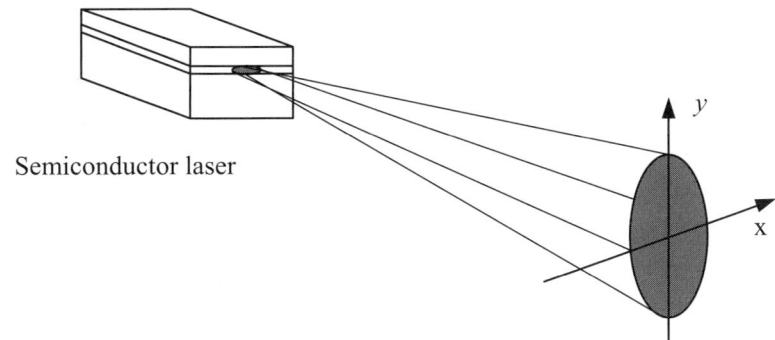

Figure 3.18. The far-field pattern from a typical single-mode laser diode.

The large divergences of the semiconductor laser beam may cause problems, for example, in coupling light, into optical fibers. An efficient method of improving coupling efficiency is the use of lenses to transform the output from the laser. Since the mode spot size of semiconductor lasers is much smaller than that of the fibers, by magnifying the output we can reduce the divergence of the beam and achieve a good coupling efficiency. For maximum efficiency, the Gaussian field profile of the semiconductor laser should be made to match that of the fiber mode. The coupling lenses can be external to the fiber or can be formed at the tip of the fiber itself by etching. With such techniques, about 50% coupling efficiency to single-mode fibers can be achieved.

Because the radiative transitions of electrons in a semiconductor take place between the valence band and the conduction band, the spontaneous spectral bandwidth of the semiconductor is larger than that of an isolated atomic system. Thus, even though the optical resonant cavity of the semiconductor laser is relatively short, there still can be a number of frequencies oscillating simultaneously. Therefore, if there is no specific treatment, semiconductor lasers normally oscillate in multiple longitudinal modes.

There are a number of ways to achieve single-longitudinal-mode oscillation. The simple one, of course, is increasing the longitudinal mode spacing by shortening the resonant cavity in order that only one longitudinal mode can be within the gain spectrum. However, shortening the cavity may create problems in handling, and also, because the volume of the gain medium is restricted, the output power is limited.

The more efficient and reliable method to achieve single-longitudinal-mode oscillation is to introduce a component or mechanism into the laser cavity that causes a loss for all longitudinal modes except one. In this way, only one of the longitudinal modes for which the gain exceeds the loss would be able to oscillate. The commonly used techniques include employing an extra

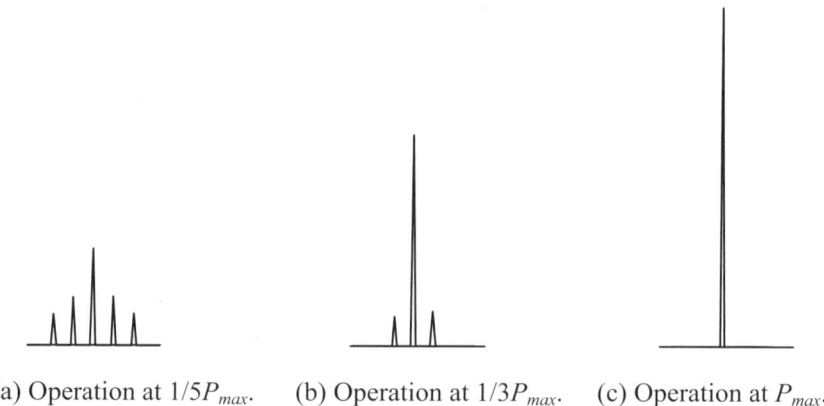

(a) Operation at $1/5P_{max}$.   (b) Operation at $1/3P_{max}$.   (c) Operation at P_{max}.

Figure 3.19. Output spectra of a single-mode semiconductor laser.

cavity (coupled cavity), using gratings (distributed Bragg reflector (DBR)) at the end sections of the optical cavity, or even integrating the gratings over the entire cavity region (distributed feedback (DFB) structure).

Figure 3.19 shows the typical output spectra from the single-mode semiconductor laser oscillating below threshold and above threshold. The frequency bandwidth of single-longitudinal-mode semiconductor lasers is typically in the range of 5–20 MHz. For single-longitudinal-mode QW semiconductor lasers, the frequency bandwidth can be significantly reduced and is typically in the range of 0.9–1.3 MHz.

The light emitted from semiconductor lasers is mostly polarized, and the electric field vector is in the plane of the *p-n* junction. The polarization ratio of the broad-area semiconductor lasers is typically 30:1 or greater, while the polarization ratio of the dielectric waveguide semiconductor lasers is in the range from 50:1 to 100:1.

3.3.5 Frequency Modulation of Semiconductor Lasers

The most attractive feature of the semiconductor laser is that its optical frequency can be directly modulated simply by modulating the drive current. The physical background behind this is that the change of the drive current alters the carrier concentration in the active medium in the laser cavity, which in turn changes the refractive index of the semiconductor and the laser oscillation frequency. For most commercial single-mode semiconductor lasers, the optical frequency modulation excursion can be up to 100 GHz without phase interruption and frequency hopping. This value is approximately equal to the frequency space between the individual longitudinal modes in a multiple-mode semiconductor laser.

As indicated earlier, the modulation bandwidth of the index-modulation method is limited by the Q factor of the laser cavity. Semiconductor lasers generally have a large Q factor. The photon lifetime (the average time that a photon spends inside the cavity before either escaping from the cavity or being absorbed or scattered) is typically about 2 ps. Hence, semiconductor lasers usually can be modulated at a high frequency (up to 10 GHz). However, modulating the frequency of a semiconductor laser usually broadens its spectral bandwidth in the range of 0.01–0.2 nm. If the modulation frequency is too high, a single-mode semiconductor laser may even oscillate in a multiple mode. Therefore, for optical FMCW interference, the modulation frequency of semiconductor lasers normally is in the range of $1–10^4$ kHz.

Note that the frequency modulation of a semiconductor laser is always accompanied by an intensity modulation. This is because the change of the carrier concentration in the laser cavity also changes the gain coefficient. Figure 3.20 shows the waveforms of the real signals from a sawtooth-wave Mach-Zehnder FMCW interferometer (see Section 6.3) when the drive current is modulated from below the threshold to beyond the maximum. The upper trace is the waveform of the drive current, while the lower trace is the waveform of the beat signal. The carrot-shaped waveform of the beat signal demonstrates that the optical property of the laser beam is not identical over the whole modulation period.

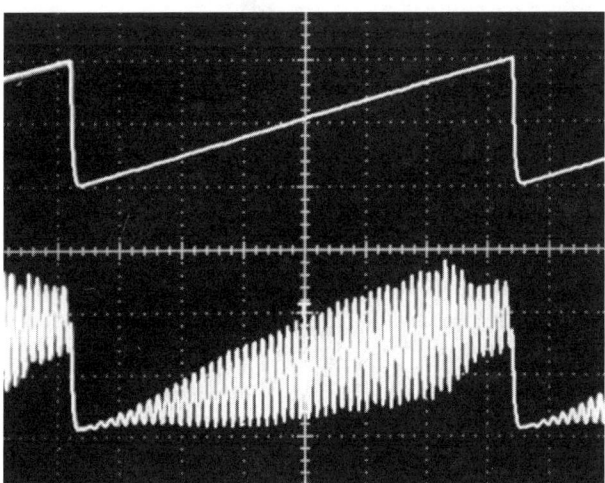

Figure 3.20. Waveforms of the real signals from a sawtooth-wave Mach-Zehnder FMCW interferometer when the semiconductor laser is driven by a widely modulated current. The upper trace is the waveform of the drive current, the lower trace is the waveform of the beat signal. (Photo by the author.)

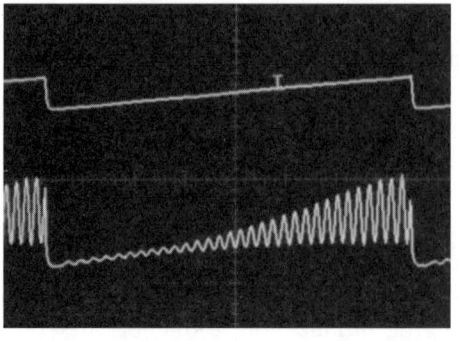

(a) Waveform of the beat signal
with a poor contrast.

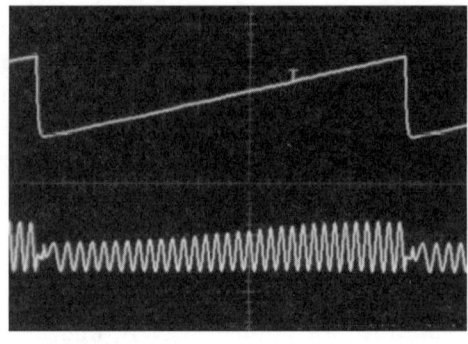

(b) Waveform of the beat signal
with an acceptable quality.

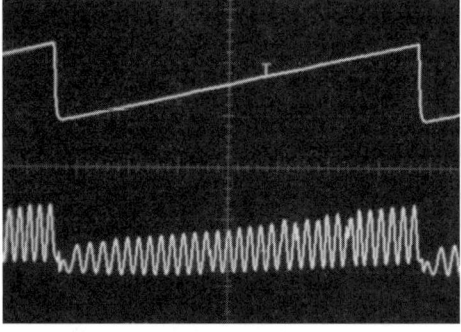

(c) Waveform of the beat signal
with a phase interruption.

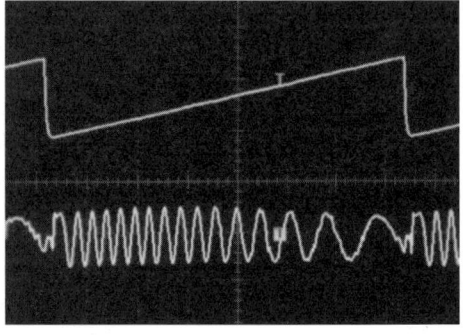

(d) Waveform of the beat signal
with a frequency reduction.

Figure 3.21. Waveforms of the beat signal under different conditions. (Photo by the author.)

Figure 3.21 shows the waveforms of the beat signal under different conditions. If the driving current is below the threshold, because the light is mainly produced by spontaneous emission, the light basically is incoherent and thus the contrast of the beat signal is almost equal to zero, as shown in Figure 3.21(a). If the driving current is above the threshold, since the radiation process principally is dominated by stimulation emission, the spectral bandwidth of the semiconductor laser gets narrower, and the contrast of the beat signal becomes bigger, as shown in Figure 3.21(b). If the driving current exceeds the maximum, a phase interruption or frequency hopping in the semiconductor laser will occur, causing a phase interruption in the beat signal, as shown in Figure 3.21(c). If the driving current approaches the catastrophic limit, a serious nonlinear frequency-current response will occur, causing the frequency of the beat signal to decrease rapidly, as shown in Figure 3.21(d).

Note that, in this situation, the semiconductor laser will probably be permanently damaged. Hence, in practice, the ideal drive current for a semiconductor laser should be in the middle region 80–90% of the range between the threshold and the maximum.

3.3.6 Driving Circuits for Semiconductor Lasers

Semiconductor lasers are current-operated devices. They must always be operated under a forward bias with a resistor in series or driven by a current source. Figure 3.22 shows a simple driving circuit for semiconductor lasers. A current-restriction resistor R is connected with a semiconductor laser in series to confine current through the laser. The value of R is determined by

$$R = \frac{V - V_f}{I},$$
(3-54)

where V_f is the laser forward voltage drop (typically about 2 V), and I is the required current (typically between 40 mA and 50 mA).

This circuit is very simple and works well. However, caution must be exercised when you test the circuit because touching the components may change the state of the circuit and destroy the semiconductor laser instantly.

Figure 3.23 shows a more practical driving circuit, which is based on a transconductance amplifier (i.e., a voltage-to-current converter), where R_L is a load resistor and *OP* is a power operational amplifier. The transconductance gain A of the circuit is given by

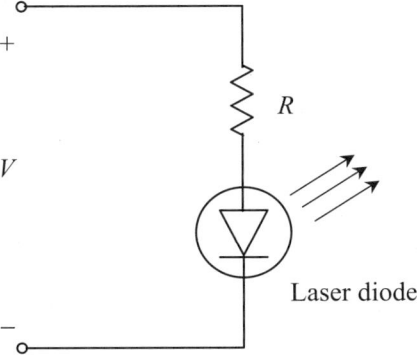

Figure 3.22. A simple driving circuit for semiconductor lasers.

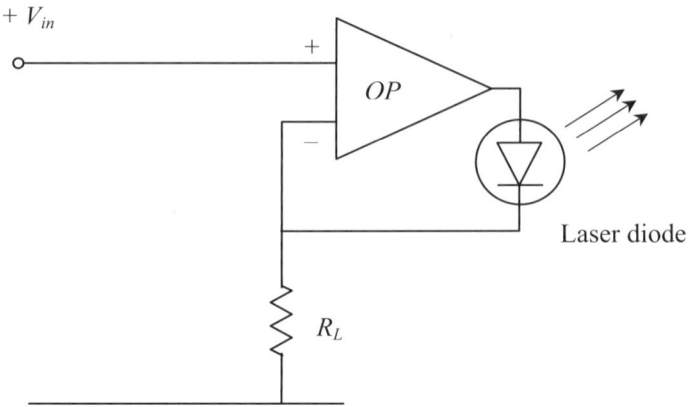

Figure 3.23. A practical driving circuit for semiconductor lasers.

$$A = \frac{I_{out}}{V_{in}}$$

$$= \frac{1}{R_L} \, ,$$

(3-55)

where I_{out} is the output current and V_{in} is the input voltage. This driving circuit is very stable and safe.

3.3.7 Laser Noise, Frequency Drift, and Feedback Light Effect

The optical wave emitted by any real light source always has some differences from the ideal regular waveform. The variation in amplitude is called the amplitude noise (or intensity noise), while the fluctuation in phase is called the phase noise (or frequency noise). Semiconductor lasers are relatively low-noise sources. However, if the frequency of a semiconductor laser is modulated, the noise generally becomes remarkable.

There are a number of noise sources in a frequency-modulated semiconductor laser. In addition to the imperfection of the drive current, other important noise sources include spontaneous emission, the photon quantum mechanism, and photon lifetime in the resonant cavity. Spontaneous emission in a semiconductor laser not only provides an intensity background but also produces an incoherent noise. The quantum nature and random lifetime of photons in the resonant cavity make the optical frequency vary in steps irregularly, rather than smoothly and uniformly. Moreover, the nonlinear frequency modulation response, intensity modulation, and spectral bandwidth variation also affect the intensity and contrast of the beat signal. However, these effects generally are regular, and therefore they might be compensated in signal processing.

One of the important characteristics of semiconductor lasers is that the light intensity and the optical frequency are strongly dependent on the surrounding temperature. A rise in temperature reduces the output power and shifts the laser spectrum to the long-wavelength direction. Figure 3.24 shows a typical output power variation with the drive current at various temperatures. Notice that the threshold current depends critically on temperature. Typically, the increase in I_{th} is 0.6–1%/°C for GaAlAs lasers and 1.2–2%/°C for InGaAsP lasers.

The frequency drift of semiconductor lasers due to temperature is about 25 GHz/°C. This redundant temperature sensitivity makes the phase measurement difficult. For instance, a 1°C change in the semiconductor laser temperature when using an interferometer with a 10 cm path imbalance results in a 50 rad drift in the phase of the beat signal. For this reason, in practice, semiconductor lasers normally are mounted on a copper heat sink and a Peltier heating element is used to stabilize the laser temperature within 0.01 °C of the selected operating temperature, which results in a phase drift of the output signal of less than 0.5 rad. For a high-precision phase measurement, a phase-drift compensation system is necessary. The simplest one uses an additional optical interferometer system with a constant path imbalance to measure the phase drift and then to rectify the measurement data.

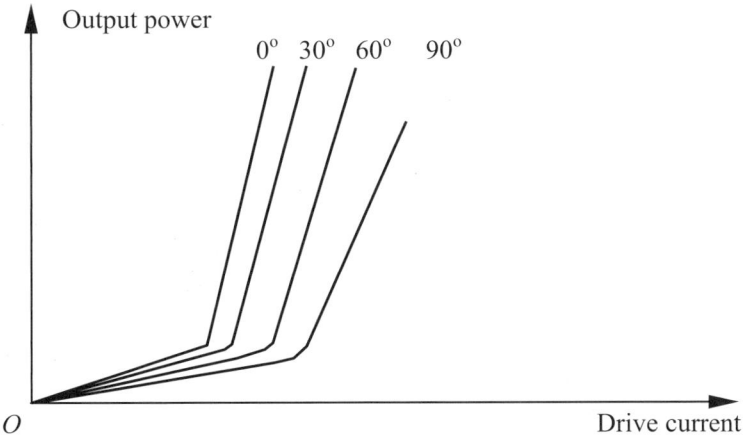

Figure 3.24. Output power variation with the drive current at various temperatures (in degree C).

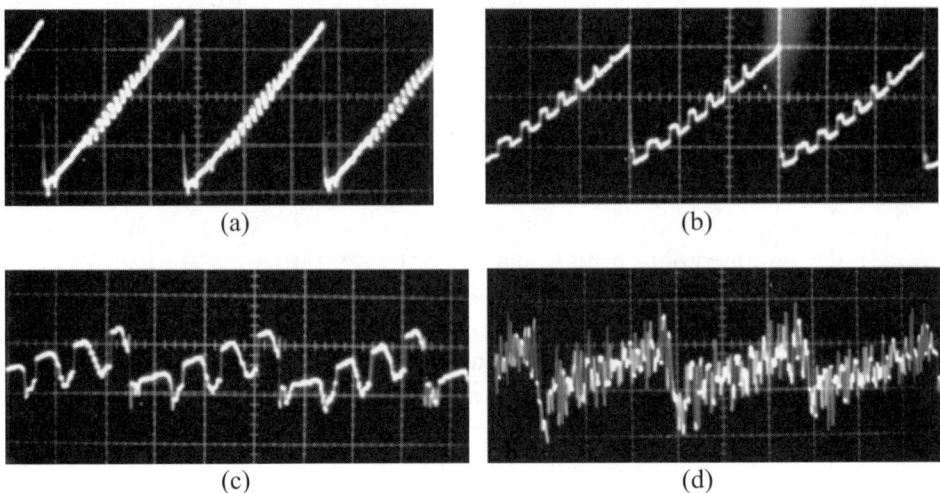

Figure 3.25. Feedback-light effects on a semiconductor laser. (Photo by the author.)

The feedback-light effect is another critical issue for semiconductor lasers. The feedback light from the optical system, particularly from the fiber-optic system, can significantly affect the property of the laser, cause a much higher level of noise, and even create an awful result. Figure 3.25 shows some intensity waveforms of a single-mode semiconductor laser, as the laser beam is coupled into a single-mode optical fiber by using ordinary microscope objective lenses. From Figures 3.25(a) to (d), the optical axes of the semiconductor laser, lenses, and optical fiber gradually get closer, the feedback light from the lens surfaces and fiber end gradually gets bigger, and thus the laser noise gradually becomes more serious. Slight misalignment of the optical axes of the components can reduce the feedback light, but this will reduce the light-coupling efficiency. Moreover, extending the fiber length to exceed the coherence length of the laser can also degrade the effect of the feedback light.

Hence, in practice, the feedback light should be restricted at an extremely low level. In general, for transmissive optical systems, all the reflecting surfaces should be coated with an antireflection film; for reflectometric optical systems, an additional optical isolator is necessary.

Chapter 4

Optical Detectors for Optical Frequency-Modulated Continuous-Wave Interference

Optical detectors are essential components for optical FMCW interference. They are also one of the crucial elements that dictate the overall performance of an optical FMCW interferometric system. In this chapter, we are going to briefly introduce some basic knowledge about optical detectors, analyze the most commonly used semiconductor photodiodes (including PN photodiodes, PIN photodiodes and avalanche photodiodes), and discuss photodiode biasing, photocurrent amplification, and noise sources.

4.1 Introduction to Optical Detectors

Optical detectors (or *photodetectors*) are devices that convert light signals into electrical signals. There can be two categories of optical detectors: thermal detectors and quantum detectors. *Thermal detectors* (such as thermocouples, bolometers, and pyroelectric detectors) respond in proportion to incident light energy (i.e., heat) and usually have a slow response time (generally more than 1 ms).

Quantum detectors (or *photon detectors*) respond in proportion to the incident photon rate and usually have a fast response time (up to 0.1 ns). Photon detectors can be further grouped into photoemissive detectors (such as vacuum photodiodes and photomultiplier tubes), photoconductive detectors (such as photoresistors), photovoltaic detectors (such as photocells, photodiodes, and phototransistors), and photoelectromagnetic detectors.

The performance of an optical detector usually is described by the following parameters.

(1) Sensitivity: The smallest amount of optical radiation that the detector can sense, usually represented by noise-equivalent power (*NEP*). The *NEP* is defined as the incident light power at a particular wavelength or specified spec-

tral bandwidth required to produce a photocurrent equal to the root-mean-square (rms) value of the noise current within a specified bandwidth. For convenience, in practice, we usually perform the measurement under high-power illumination and use Equation (4-1) to compute the *NEP*,

$$NEP = \frac{P_s}{V_s / V_n},$$

(4-1)

where P_s is the incident light power, V_s is the output signal voltage, V_n is the output noise voltage, and V_s/V_n is called the signal-to-noise ratio.

(2) Responsivity ($\Re$): The ratio between the output signal (voltage or current) and the incident light power. The responsivity generally is wavelength-dependent, and the relationship between responsivity and wavelength is called the spectrum responsivity, represented by $\Re(\lambda)$.

(3) Time constant (τ)/chopping frequency (f_c): A measure of the response speed of an optical detector. For most photodetectors, the relationship between the responsivity $\Re$ and the modulation frequency f of the incident light signal can be represented by

$$\Re(f) = \frac{\Re_0}{\sqrt{1 + 4\pi^2 f^2 \tau^2}},$$

(4-2)

where $\Re_0$ is the responsivity when f equals zero, and τ is a constant (called the *time constant*). The chopping frequency f_c is defined as the frequency at which the responsivity has fallen to 0.707 of its maximum value. The chopping frequency f_c and the time constant τ are related by

$$f_c = \frac{1}{2\pi\tau}.$$

(4-3)

(4) Quantum efficiency (η): The ratio of the number of countable output events (such as electron-hole pairs in semiconductor photodiodes) to the number of incident photons, usually represented as a percentage value.

(5) Dark current: The output current that flows when input optical radiation is absent or negligible. Notice that, although this current can be subtracted out by using an electric circuit, the shot noise on the dark current can become the foremost noise source.

Moreover, there are some other parameters traditionally employed to assess the noise performance of optical detectors. The ones commonly used are detectivity (D) and specific detectivity (D^*). The detectivity is the reciprocal of the noise-equivalent power. It can also be expressed as the rms signal-to-noise ratio per unit of the light power incident on the detector,

$$D = \frac{1}{NEP}$$
$$= \frac{V_s / V_n}{P_s} .$$

(4-4)

The specific detectivity is a normalization of the reciprocal of the noise-equivalent power to take into account the area and electric-bandwidth dependence,

$$D^* = \frac{\sqrt{A\Delta f}}{NEP}$$
$$= \sqrt{A\Delta f} D ,$$

(4-5)

where A is the sensing area of the photodetector and Δf is the frequency bandwidth. The D^* parameters can be used to compare optical detectors of different types because the detector with a greater value of D^* is a better detector when the other parameters are identical.

As to the optical detector used for optical FMCW interference, the essential requirements could be the six that follow.

(1) High sensitivity: The sensitivity of the detector should be high, so that a weak beat signal can be detected.

(2) Short response time: Because the beat frequency of optical FMCW interference can be very high (up to 1 GHz), the response time of the detector should be very short.

(3) Linear responsivity: The detector should be linear in response to the light intensity, so that an optical beat signal can be converted into an electrical beat signal without distortion.

(4) High quantum efficiency: The detector should produce a maximum electrical signal for a given amount of light power.

(5) Low noise: The detector should produce a low dark current and little shot noise.

(6) Low bias voltage: The detector should not require an excessive bias voltage or current.

In addition to the above, the stability, reliability, physical size, and cost of the detector are also factors to consider for practical applications.

Thus far, scmiconductor photodiodes are believed to be the best solution for optical detection of optical FMCW interference since they have a number of desirable properties, such as high sensitivity, short response time, acceptable linear responsivity, low noise, sturdiness, small size, and relatively low cost. Hence, in the following sections, we will concentrate our attention on semiconductor photodiodes.

4.2 Semiconductor Photodiodes

The operating principle of semiconductor photodiodes is based on optical absorption. When light is incident on a semiconductor, the light may be absorbed if the energy hv of a photon of the incident light beam is greater than the band gap of the semiconductor. The electrons of molecules may overcome the nuclear attraction and jump from the valence band to the conduction band, leading to the generation of electron-hole pairs. If an electric field is applied across the semiconductor, the photo-generated electron-hole pairs will be swept away, creating an electric current through the external circuit.

Let E_g represent the band gap between the valence band and the conduction band in a semiconductor. The maximum wavelength of absorption (or *cutoff wavelength*) λ_c is determined by

$$\lambda_c = \frac{hc}{E_g},$$

(4-6)

where h is the Planck constant and c is the speed of light in free space.

Substituting the values of h and c, the cutoff wavelength can be simplified as

$$\lambda_c \approx \frac{1.24}{E_g},$$

(4-7)

where λ_c is in micrometers and E_g is in eV.

For instance, most semiconductor photodiodes are made from indirect band-gap semiconductors, such as silicon and germanium, or from compound semiconductors, such as InGaAs. The band-gap energies are 1.11 eV for silicon, 0.67 eV for germanium, and 0.75 eV for InGaAs. Accordingly, the corresponding cutoff wavelengths are 1.12 μm for silicon, 1.85 μm for germanium, and 1.65 μm for InGaAs.

4.2.1 PN Photodiodes

The early semiconductor photodiode was made simply of a *p-n* junction, as shown in Figure 4.1(a), in which light absorption mostly takes place in the depletion region. The current-voltage characteristic of the PN photodiode is shown in Figure 4.1(b).

In practice, the PN photodiode can work in two different detection modes: photovoltaic and photoconductive. In the *photovoltaic mode*, the diode is under zero bias, and the electrons and holes generated by absorption of the incident light are collected at either end of the junction, resulting in an electric potential difference.

In the *photoconductive mode*, the diode is under reverse bias (for example, at the point Q in Figure 4.1(b)), and the electron-hole pairs generated by light absorption are separated by the high electric field in the depletion region, leading to a current in the external circuit. Photodiodes operating in the photoconductive mode have a faster response time because the carriers are swept away by the electric field. Hence, we will restrict our attention to this mode of operation.

The absorption of optical radiation in the semiconductor material can be described by

$$P = P_0(1 - e^{-ax}),$$
(4-8)

where P is the light power absorbed over distance x, P_0 is the incident light power, and a is the wavelength-dependent absorption coefficient. Considering the reflection on the air-semiconductor interface, the light power P absorbed in a distance x will be

$$P = P_0(1 - R)(1 - e^{-ax}),$$
(4-9)

where R is the reflectivity of the air-semiconductor interface. If v is the frequency of the incident light, the number of photons absorbed per unit time N_p will be

$$N_p = \frac{P_0(1 - R)(1 - e^{-ax})}{hv}.$$
(4-10)

Since each absorbed photon generates an electron-hole pair, Equation (4-10) also gives the number of electron-hole pairs generated per unit time. Assuming that only a fraction ζ of the electron-hole pairs contribute to the photocurrent (the remainder having been lost due to recombination), the photocurrent I will be

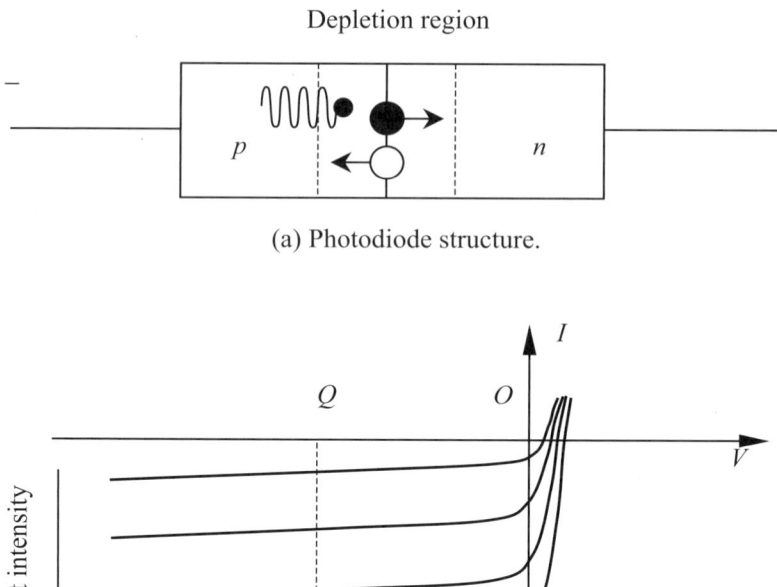

(a) Photodiode structure.

(b) Current-voltage characteristic.

Figure 4.1. PN photodiode.

$$I = \frac{e\zeta P_0 (1 - R)(1 - e^{-ax})}{h\nu} ,$$ (4-11)

where e is the magnitude of the electronic charge.

For semiconductor photodiodes, the quantum efficiency η is defined as the ratio of the number of electron-hole pairs generated to the number of incident photons. Therefore,

$$\eta = \frac{I/e}{P_0/h\nu}$$

$$= \zeta(1 - R)(1 - e^{-ax}) .$$ (4-12)

The responsivity $\Re$ is defined as the photocurrent generated per unit light power. Thus,

$$\Re = \frac{I}{P_0}$$

$$= \frac{e\zeta(1-R)(1-e^{-ax})}{h\nu}$$

$$= \frac{e\eta}{h\nu} \qquad\qquad (4\text{-}13)$$

For the silicon photodiode, for instance, the typical value of the responsivity is 0.5 A/W.

The response speed of a photodetector is primarily dependent on the transit time of the photo-generated carriers through the depletion region. The transit time τ_t of carriers across the depletion region of width w is given by

$$\tau_t = \frac{w}{v_d} , \qquad\qquad (4\text{-}14)$$

where v_d is the carrier drift velocity. Apparently, the smaller the w, the shorter the transit time τ_t will be. However, small w is obstructive for achieving larger quantum efficiency.

Apart from the transit time limitation, the photodiode capacitance also plays a significant role in response speed. The junction capacitance C_d of the photodiode can be written as

$$C_d = \frac{\varepsilon A}{w} , \qquad\qquad (4\text{-}15)$$

where ε is the permittivity of the semiconductor, and A is the p-n junction area.. The time constant τ of the photodiode can be determined by

$$\tau = R_L C_d , \qquad\qquad (4\text{-}16)$$

where R_L is the load resistance. The frequency bandwidth of the photodiode Δf can be determined by

$$\Delta f = \frac{1}{2\pi\tau} . \qquad\qquad (4\text{-}17)$$

Obviously, to achieve a small rise time, the photodiode should have a small A, large w, and small R_L. For instance, if the junction capacitance C_d of a silicon photodiode is 5 pF and the load resistance R_L equals 1000 Ω, then $\tau = 5$ ns and $\Delta f = 32$ MHz.

4.2.2 PIN Photodiodes

A PIN photodiode consists of a p-type semiconductor material, an n-type semiconductor material, and an intrinsic (or i-type) semiconductor material sandwiched between the p and n regions, as shown in Figure 4.2. Because the i region has no free charges, its resistance is relatively high, and hence most of the voltage across the diode appears across the intrinsic region. In addition, the i region is usually much wider than the p and n regions so that incoming photons have a greater probability of absorption in the i region than in the p region or n region.

Due to the high electric field in the i region, any electron-hole pair generated in this region is immediately swept away by the field. Hence, the PIN photodiode generally has a faster response. For instance, the time constant of the silicon PIN photodiode can be of a small value up to 0.5 ns. On the contrary, in a PN photodiode, the electron-hole pairs generated in the p and n regions have to first diffuse into the depletion region before being swept away, and these electron-hole pairs may recombine, resulting in a reduction in current.

The width of the intrinsic region cannot be made too large because the carriers would take a longer time to drift to the terminals and thus reduce the speed of response. For indirect-band-gap semiconductors, such as silicon and germanium, the widths are typically 20–50 µm, whereas for direct-band-gap semiconductors, such as InGaAs, the widths are typically 3–5 µm.

Figure 4.3 shows the responsivities and quantum efficiency of silicon, germanium, and InGaAs PIN photodiodes as a function of wavelength [58]. Obviously, silicon is ideal for detection in the region of 850 nm, and InGaAsis the preferred detector in the 1.3 µm and 1.55 µm wavelength regions. Notice

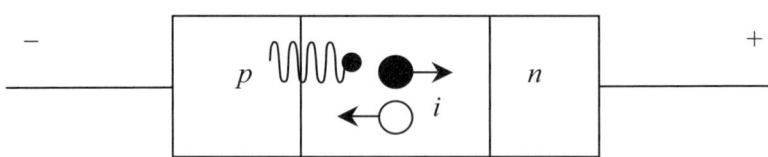

Figure 4.2. Structure of the PIN photodiode.

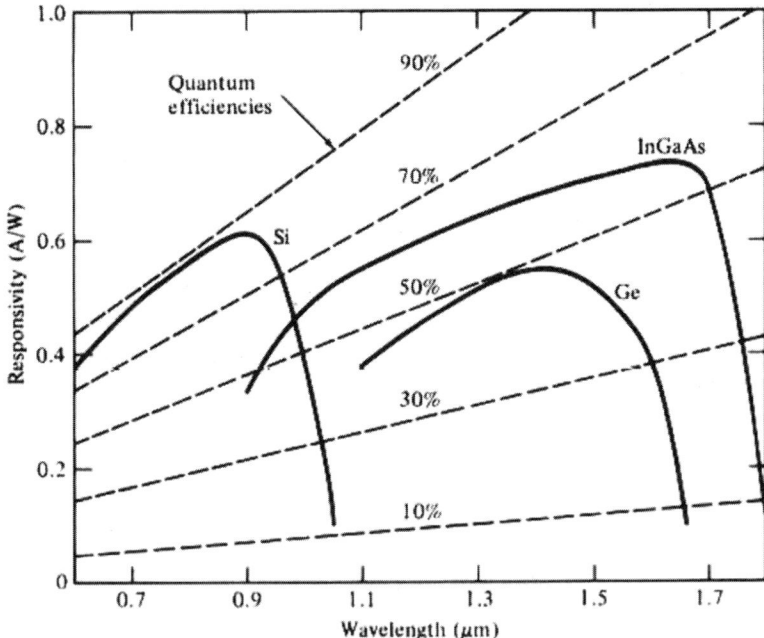

Figure 4.3. Comparison of the responsivity and quantum efficiency as a function of wavelength for PIN photodiodes constructed of different materials. (From Gerd Keiser, *Optical Fiber Communications, 2nd edition*, McGraw-Hill Inc ©1991, reprinted with permission of McGraw-Hill Education.)

that the long-wavelength cutoff of a photodetector is caused by the fact that the energy of the incident photons is less than the band gap, while the short-wavelength cutoff of a photodetector is caused by a very large value of the absorption coefficient a. This large absorption coefficient results in absorption very close to the photodetector surface, where the electron-hole recombination time is very short. Thus, the photo-generated electron-hole pairs recombine within the detector itself rather than contributing to the current in the external circuit.

4.2.3 Avalanche Photodiodes (APD)

The avalanche photodiode (APD) has an internal current gain, which is achieved by applying a large reverse voltage bias. The absorption of an incident photon first produces an electron-hole pair just as in a PIN photodiode. The large electric field in the depletion region causes the charges to accelerate rapidly. The fast-moving charges can give part of their energy to an electron in the valence band and excite it into the conduction band, which results in the

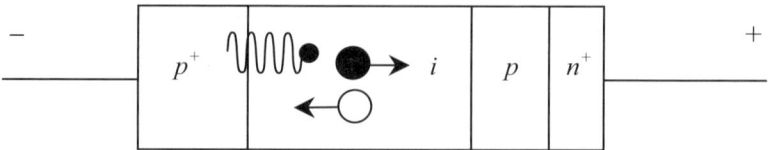

Figure 4.4. Structure of the avalanche photodiode.

creation of an additional electron-hole pair. The newly produced charges in turn can further accelerate and create more electron-hole pairs. Eventually, an avalanche multiplication of the carriers occurs.

Unlike the PIN photodiode, the APD generally has an additional p-type (or n-type) layer between the intrinsic and the highly doped n-type regions, as shown in Figure 4.4. The electron-hole pairs are still generated in the i region, but the avalanche multiplication takes place in the p region. The benefit of this extra layer is that the avalanche multiplication happens just to one type of charge carrier, so the shot noise can be largely reduced.

The multiplication factor M of the APD can be expressed by

$$M = \frac{1}{1 - \left(\dfrac{V}{V_b}\right)^n},$$

$$(4\text{-}18)$$

where V is the reverse bias applied, V_b is the breakdown voltage, and n is a constant (varying between 3 and 6, depending on the semiconductor material and the substrate type). When $V = V_b$, $M = \infty$, the photodiode will break down. The responsivity $\Re_{APD}$ of the APD can be written as

$$\Re_{APD} = M\Re = \frac{Me\eta}{h\nu},$$

$$(4\text{-}19)$$

where $\Re$ is the responsivity of the corresponding PIN photodiode.

The large reverse voltage bias can shorten the charge-collision time and reduce junction capacitance. Therefore, the response time of the APD generally is shorter than that of the PIN photodiode. For instance, the time constant for the silicon APD can be close to 0.1 ns. On the other hand, however, the large reverse bias magnifies the shot noise. Moreover, because the multiplication factor M itself fluctuates around the mean, the APD has an additional noise (named excess noise). Generally, the multiplication factor of the shot-noise

Table 4.1. Characteristics of PIN and avalanche photodiodes. (From A. Ghatak and K. Thyagarajan, *Introduction to Fiber Optics*, Cambridge University Press ©1998, reprinted with permission of Cambridge University Press.)

Parameter	Silicon		Germanium		InGaAs	
	PIN	APD	PIN	APD	PIN	APD
Wavelength range (nm)	400–1100		800–1800		900–1700	
Peak (nm)	900	830	1550	1300	1300 (1550)	1300 (1550)
Responsivity $\mathfrak{R}$ (A/W)	0.6	77–130	0.65–0.7	3–28	0.63–0.8 (0.75–0.97)	
Quantum efficiency (%)	65–99	77	50–55	55–75	60–70	60–70
Gain (M)	1	150–250	1	5–40	1	10–30
Excess-noise factor (x)	—	0.3–0.5	—	0.95–1	—	0.7
Bias voltage (−V)	45–100	220	6–10	20–35	5	<30
Dark current (nA)	1–10	0.1–1.0	50–500	10–500	1–20	1–5
Capacitance (pF)	1.2–3	1.3 –2	2–5	2–5	0.5–2	0.5
Rise time (ns)	0.5–1	0.1–2	0.1–0.5	0.5–0.8	0.06–0.5	0.1–0.5

current of an APD can be written as $M^{1+\sqrt{x}}$, where x is called the excess-noise factor. Hence, in practice, the multiplication factor and noise should balance in order to achieve the best operation. Table 4.1 shows some important characteristic properties of silicon, germanium, and InGaAs PIN and avalanche photodiodes [93].

4.3 Photodiode Biasing and Signal Amplification

As indicated earlier, semiconductor photodiodes operate under a reverse bias and produce a current signal. This current signal normally is relatively weak and is required to be amplified and transformed into a voltage signal for further processing. In practice, photodiode basing, photocurrent-voltage transformation, and signal preamplification usually are completed by using a single electric circuit.

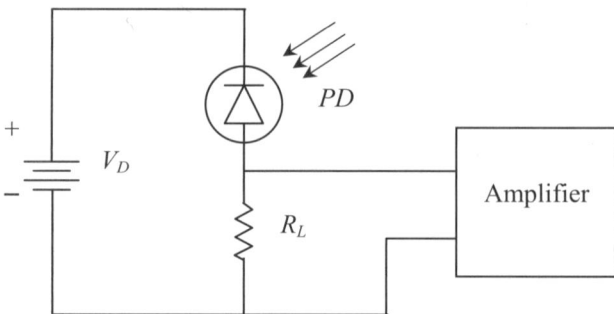

Figure 4.5. Photodiode circuit with a load resistor and voltage amplifier.

There are two approaches to accomplish the three tasks above. The first approach is to change the current signal into a voltage signal by connecting the photodiode with a load resistor in series and then amplify the voltage signal across the load resistor by using a voltage amplifier, as shown in Figure 4.5, where V_D is a dc voltage source whose value depends on the type of photodiode, PD is a photodiode, and R_L is the load resistor that provides a direct-current path back to the electric source and whose value dominates the frequency bandwidth of the detector (see Equation (4-16)).

Figure 4.6 shows a practical circuit for the silicon PIN photodiode, where V_D is a voltage source, R_1 and C_1 are the filtering components used for preventing the output signal to feed back to the input end through the power line, PD is a photodiode, R_L is a load resistor, and OP is a low-noise operational amplifier. The transresistance gain A of the circuit is given by

$$A = \frac{V_{out}}{I_{in}}$$
$$= R_L\beta , \tag{4-20}$$

where V_{out} is the output voltage, I_{in} is the input photocurrent, and β is the open-loop gain of the operational amplifier.

In the second approach, the photodiode is treated as a current source, and a transresistance amplifier (i.e., a current-to-voltage converter) is used to convert and amplify the photocurrent simultaneously. The second approach may offer the potential for a lower noise and wider bandwidth than the configuration of a load resistor and a voltage amplifier.

Figure 4.7 shows a typical transresistance amplifier configuration, where $FET\ OP$ is a field-effect transistor operational amplifier that is used because of

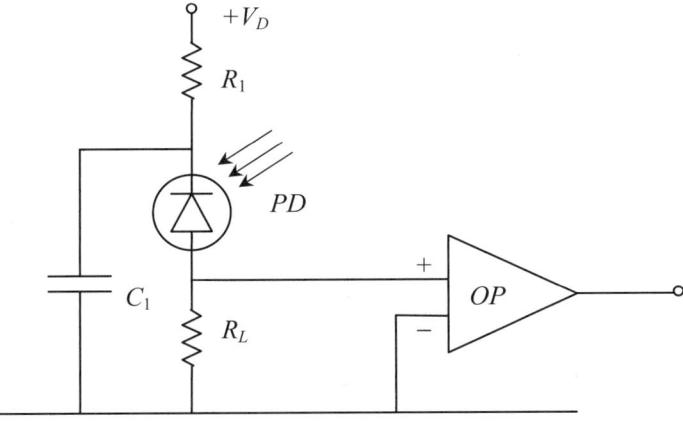

Figure 4.6. A practical circuit for the silicon PIN photodiode.

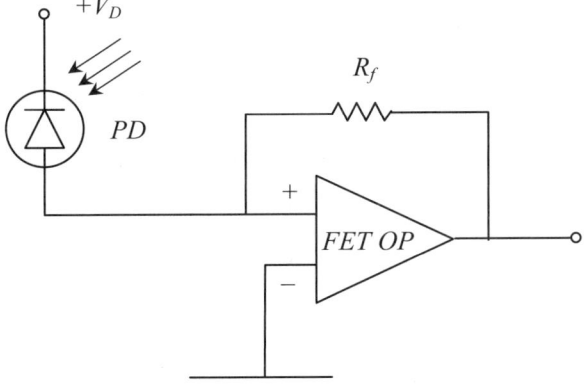

Figure 4.7. Photodiode circuit with a transresistance amplifier.

its low-input bias current, and R_f is the feedback resistor. The transresistance gain A is given by

$$A = \frac{V_{out}}{I_{in}}$$

$$= R_f \; , \qquad (4\text{-}21)$$

where V_{out} is the output voltage and I_{in} is the input photocurrent.

4.4 Noise in the Detection Process

Understanding noise sources in photodetection is crucial to achieving the best signal-to-noise operational performance. Any trivial error in the measurement may reduce the accuracy significantly. For instance, improperly choosing the load resistor for a photodiode may increase the current noise by several orders of magnitude.

Semiconductor photodiodes, like other semiconductor elements, have a number of noise sources. The important ones are discussed next.

(1) Shot noise: Shot noise arises due to the fact that the electric current consists of a stream of discrete charges and that the charge carries in a semi-conductor material are randomly generated. Thus, even when a photodiode is illuminated by constant light power, the photocurrent will fluctuate randomly around an average value determined by the average light power.

Shot noise is a kind of white noise, and its average value is equal to zero. Generally, we use the root-mean-square I_{shot} of the shot noise current to represent the intensity of the shot noise. According to semiconductor physics, I_{shot} equals

$$I_{shot} = \sqrt{2eI\Delta f}\ ,$$

(4-22)

where e is the magnitude of the electronic charge, I is the average photocurrent generated by the photodiode, and Δf is the bandwidth over which the noise is being considered. Since the photocurrent I itself depends on the incident light power, the shot noise increases with the rise in power of the incident light.

Notice that any photodiode has a dark current, which emerges from the thermally generated carriers. Considering the dark current, Equation (4-22) becomes

$$I_{shot} = \sqrt{2e(I + I_d)\Delta f}\ ,$$

(4-23)

where I_d is the dark current.

(2) Thermal noise (also called Johnson noise or Nyquist noise): Thermal noise arises because the random motion of charge carries in a resistive element (such as the load resistor in a photodiode circuit) at thermal equilibrium generates a random electric voltage across the element. Since the electron motion is random, the average of this current is zero. The root-mean-square value of the thermal noise $V_{thermal}$ can be written as

$$V_{thermal} = \sqrt{4k_B TR\Delta f}\ ,$$

(4-24)

where k_B is Boltzmann's constant, T is the absolute temperature, R is the resistance, and Δf is the frequency bandwidth. Notice that the thermal noise current increases as R decreases, and therefore a small-value load resistor should be avoided. Unfortunately, a small load resistor is required to maintain a wide frequency response.

(3) $1/f$ Noise: The mechanism involved in this type of noise is not well-understood. It is generally believed that the voltage across a resistor carrying a constant current will fluctuate because the resistance of the material used in the resistor varies. The magnitude of the resistance fluctuation depends on the material used. Carbon composition resistors are the worst, metal-film resistors are better, and wire-wound resistors provider the lowest $1/f$ noise. The rms value of this noise source for a resistor $V_{1/f}$ is given by

$$V_{1/f} = IR\sqrt{\frac{A\Delta f}{f}},$$

(4-25)

where A is a dimensionless constant (for example, A is about 10^{-11} for carbon), R is the resistance, f is the central frequency, and Δf is the frequency bandwidth.

In general, the noise of semiconductor photodiodes is much smaller than that of the following amplifiers. The entire noise of the optical measurement should include optical noise, photodiode noise, and amplifier noise.

Chapter 5

Coherence Theory of Optical Frequency-Modulated Continuous-Wave Interference

In the preceding chapters, all optical waves were assumed to have a single regular continuous waveform. Actually, the optical wave emitted from any real optical source is extremely complicated and unpredictable. Even the laser does not entirely satisfy the ideal assumption proposed previously. Therefore, it can be imagined that, if a number of real optical waves interfere, the intensity of the resulting electric field may be different from the theoretical prediction.

In this chapter, we will discuss the effect of the frequency bandwidth of optical sources on optical FMCW interference, coherence of optical FMCW waves, and the influence of the phase noise of optical sources on the optical FMCW beat signal.

5.1 Effect of the Frequency Bandwidth of Optical Sources

In physical optics, the basic assumption for the analysis is that the optical waves with different frequencies are incoherent, and the optical wave from any real optical source consists of a number of frequencies (also called the frequency bandwidth). For this reason, if two optical waves, derived from the same real optical source but traversing different paths, are recombined to interfere, the intensity of the resulting electric field should be determined by the sum of the intensities produced by all individual optical frequencies.

This theory, however, is hardly capable of describing the behavior of the optical FMCW wave and the phenomenon of optical FMCW interference because the fundamental assumption for optical FMCW interference is that the optical waves with varying frequencies are coherent.

In fact, the optical wave emitted by any real optical source can be believed to be composed of a number of incoherent components. If the frequency of the optical wave is not modulated, the frequencies of all individual components are not variable but different, and therefore these components can be identified by their frequencies (or angular frequencies), as shown in Figure 5.1(a). If the frequency of the optical wave is modulated with a specific waveform, however, the frequencies of all individual components will be modulated in the same way, and these components should be identified by their central frequencies (or central angular frequencies), as shown in Figure 5.1(b). Note that, since the composition of the optical wave does not change when the optical frequency is modulated, the central frequency bandwidth with frequency modulation should be the same as the frequency bandwidth without frequency modulation.

This hypothesis, of course, does not contradict the classical coherence theory for conventional optical waves and classical optical interference, but it can explain the coherence phenomenon in optical FMCW interference. For instance, if two optical FMCW waves derived from the same narrow-band optical source whose frequency is modulated with a sawtooth waveform but traversing different paths, are recombined to interfere, the intensity of the resulting electric field should be

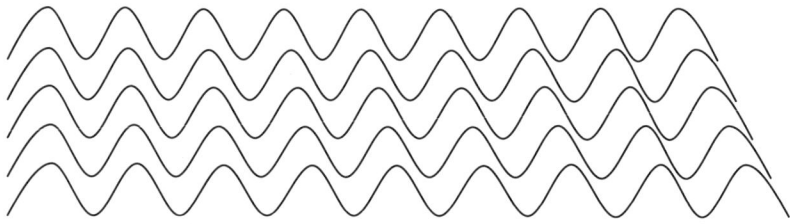

(a) A limited-band optical wave without frequency modulation.

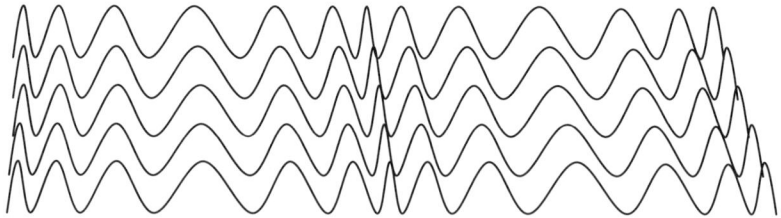

(b) A limited-band optical wave with frequency modulation.

Figure 5.1. Composition of practical optical waves.

$$I(t) = \int S(\omega_0)(i_1 + i_2)\left[1 + \frac{2\sqrt{i_1 i_2}}{i_1 + i_2}\cos(\alpha\tau t + \omega_0\tau)\right]d\omega_0$$

$$= (i_1 + i_2)\left\{\int S(\omega_0)d\omega_0 + \frac{2\sqrt{i_1 i_2}}{i_1 + i_2}\left[\int S(\omega_0)\cos(\alpha\tau t)\cos(\omega_0\tau)d\omega_0\right.\right.$$

$$\left.\left. - \int S(\omega_0)\sin(\alpha\tau t)\sin(\omega_0\tau)d\omega_0\right]\right\}$$

$$= (i_1 + i_2)\left\{F(0) + \frac{2\sqrt{i_1 i_2}}{i_1 + i_2}[\cos(\alpha\tau t)F_c(\tau) - \sin(\alpha\tau t)F_s(\tau)]\right\}$$

$$= (i_1 + i_2)\left\{F(0) + \frac{2\sqrt{i_1 i_2}}{i_2 + i_2}|F(\tau)|\left[\cos(\alpha\tau t)\frac{F_c(\tau)}{F(\tau)} + \sin(\alpha\tau t)\frac{-F_s(\tau)}{F(\tau)}\right]\right\}$$

$$= (i_1 + i_2)\left\{F(0) + \frac{2\sqrt{i_1 i_2}}{i_1 + i_2}|F(\tau)|\cos[\alpha\tau t - \theta(\tau)]\right\}$$

$$= (I_1 + I_2)\left\{1 + \frac{2\sqrt{I_1 I_2}}{I_1 + I_2}\frac{|F(\tau)|}{F(0)}\cos[\alpha\tau t - \theta(\tau)]\right\}$$

$$= I_0\{1 + V\cos[\alpha\tau t - \theta(\tau)]\} \tag{5-1}$$

where $S(\omega_0)$ is the power spectrum of the optical source, i_1 and i_2 are relative intensities of the two interfering waves corresponding to each wave component, $F_c(\tau)$ is the Fourier cosine transformation of $S(\omega_0)$, $F_s(\tau)$ is the Fourier sine transformation of $S(\omega_0)$, $F(\tau)$ is the Fourier transformation of $S(\omega_0)$, $\theta(\tau)$ is the angle of $F(\tau)$, I_1 and I_2 are intensities of the two interfering waves, I_0 is the average intensity, V is the contrast of the beat signal, the integration range formally extends to $\pm\infty$, and some quantities are given by

$$F_c(\tau) = \int S(\omega_0)\cos(\omega_0\tau)d\omega_0 \tag{5-2}$$

$$F_s(\tau) = \int S(\omega_0)\sin(\omega_0\tau)d\omega_0 \tag{5-3}$$

$$F(\tau) = \int S(\omega_0)e^{-j\omega_0\tau}d\omega_0 \tag{5-4}$$

$$F(\tau) = F_c(\tau) - jF_s(\tau) \tag{5-5}$$

$$\theta(\tau) = \arg[F(\tau)]$$

$$= \cos^{-1}\left(\frac{F_c(\tau)}{F(\tau)}\right)$$

$$= \sin^{-1}\left(\frac{-F_s(\tau)}{F(\tau)}\right),$$

(5-6)

$$I_1 = i_1 F(0),$$

(5-7)

$$I_2 = i_2 F(0),$$

(5-8)

$$I_0 = I_1 + I_2,$$

(5-9)

$$V = \frac{2\sqrt{I_1 I_2}}{I_1 + I_2} \frac{|F(\tau)|}{F(0)}.$$

(5-10)

If the power spectrum of the optical source has a distribution of rectangle function

$$S(\omega_0) = \begin{cases} 1 & |\omega_0 - \Omega_0| \le \Delta\omega_0 / 2 \\ 0 & |\omega_0 - \Omega_0| > \Delta\omega_0 / 2 \end{cases},$$

(5-11)

where $\Delta\omega_0$ is the central angular frequency bandwidth, then

$$F(\tau) = \Delta\omega_0 \, \mathrm{Sinc}\left(\frac{\Delta\omega_0}{2}\tau\right)e^{-j\Omega_0\tau},$$

(5-12)

$$|F(\tau)| = \Delta\omega_0 \left|\mathrm{Sinc}\left(\frac{\Delta\omega_0}{2}\tau\right)\right|,$$

(5-13)

$$\theta(\tau) = -\Omega_0\tau,$$

(5-14)

$$I(t) = (I_1 + I_2)\left[1 + \frac{2\sqrt{I_1 I_2}}{I_1 + I_2}\left|\mathrm{Sinc}\left(\frac{\Delta\omega_0}{2}\tau\right)\right|\cos(\alpha\tau t + \Omega_0\tau)\right], \tag{5-15}$$

Equation (5-15) shows that the beat signal produced by two practical coherent sawtooth-wave FMCW waves still is of a single frequency and its initial phase is proportional to the middle central angular frequency Ω_0, but its contrast is modified by $|\mathrm{Sinc}(\Delta\omega_0\tau/2)|$, as shown in Figure 5.2, where the Sinc function is defined by

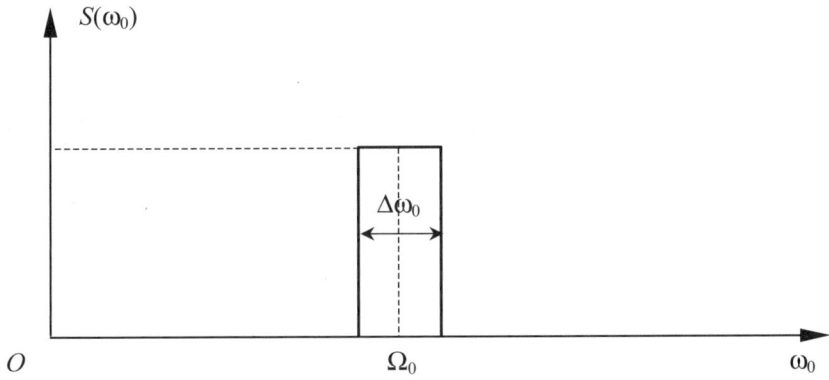

(a) Power spectrum of an optical source.

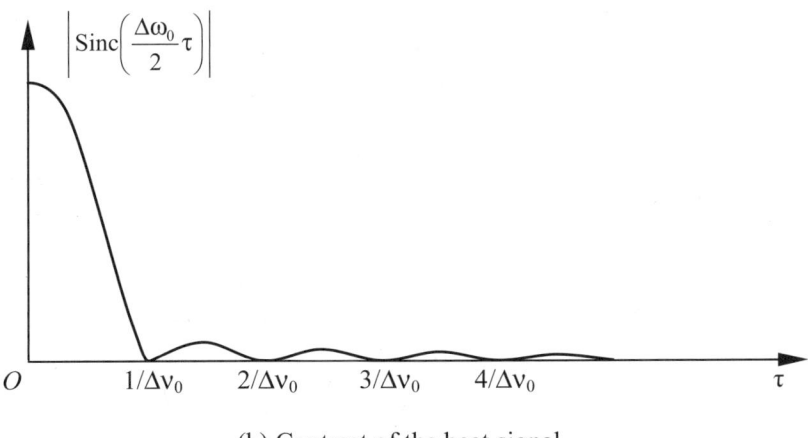

(b) Contrast of the beat signal.

Figure 5.2. Effect of the spectral bandwidth of an optical source.

$$\text{Sinc}\, x = \frac{\sin x}{x},$$

(5-16)

where x is a variable.

If we define the delay time and the optical path difference corresponding to the first zero point of the beat signal contrast as coherence time and coherence length, respectively, according to the property of $\text{Sinc}(\Delta\omega_0\tau/2)$, the coherence time t_c of the optical source will be

$$t_c = \frac{1}{\Delta v_0},$$

(5-17)

where Δv_0 is the central frequency bandwidth of the optical source ($\Delta v_0 = \Delta\omega_0/2\pi$), and the coherence length of the optical source l_c will be

$$l_c = \frac{c}{\Delta v_0},$$

(5-18)

where c is the speed of light in free space. Obviously, these relations are the same as those in the existing coherence theory.

5.2 Coherence of Optical FMCW Waves

Optical coherence, which is a complex but active research area, deals with the correlation between two or more optical waves. Optical coherence generally is divided into two aspects, temporal coherence and spatial coherence. *Temporal coherence* deals with the correlation between optical waves that are from the same point in space but emitted at different times and is directly related to the frequency bandwidth of the optical source. *Spatial coherence* deals with the correlation between optical waves that are from different points in space and is directly related to the extension of the optical source in space. For optical FMCW interference, because all practical optical FMCW waves are from single-mode lasers, and because the laser beam is coherent in space due to the nature of stimulated radiation, spatial coherence is always satisfied, and therefore only temporal coherence need be considered.

Now, let's consider the general situation. Assuming that two optical FMCW waves derived from the same quasi-monochromatic FMCW laser but propagated along different paths are recombined to interfere, the electric fields of the two waves at the observation point can be represented by $K_1 E(t)$ and $K_2 E(t-\tau)$, where K_1 and K_2 are two constants representing the variation of the amplitude due to propagation, $E(t)$ is the electric disturbance of the laser, and τ is the delay time of the second wave compared with the first one. The intensity of the resulting electric field can be written as

$$
\begin{aligned}
I(\tau,t) &= <[K_1 E(t) + K_2 E(t-\tau)]^2> \\
&= <[K_1 E(t) + K_2 E(t-\tau)][K_1 E(t) + K_2 E(t-\tau)]^*> \\
&= <K_1^2 E(t)E^*(t)> + <K_2^2 E(t-\tau)E^*(t-\tau)> \\
&\quad + <K_1 K_2 E(t)E^*(t-\tau)> + <K_1 K_2 E^*(t)E(t-\tau)> \\
&= I_1 + I_2 + K_1 K_2 \Gamma(\tau) + K_1 K_2 \Gamma^*(\tau) \\
&= I_1 + I_2 + 2K_1 K_2 \,\mathrm{Re}[\Gamma(\tau)]
\end{aligned}
\tag{5-19}
$$

where $<>$ denotes a time average over a period much longer than the optical vibration period, $\Gamma(\tau)$ is the self-coherence function of $E(t)$, and I_1 and I_2 are the intensities of the two waves. They are defined by

$$
\begin{aligned}
\Gamma(\tau) &= <E(t)E^*(t-\tau)> \\
&= \lim_{T\to\infty} \frac{1}{T} \int_0^T E(t)E^*(t-\tau)dt
\end{aligned}
\tag{5-20}
$$

$$
I_1 = K_1^2 \Gamma(0)
\tag{5-21}
$$

$$
I_2 = K_2^2 \Gamma(0)
\tag{5-22}
$$

For convenience, we usually define the normalized self-coherence function as the complex degree of temporal coherence $\gamma(\tau)$

$$
\gamma(\tau) = \frac{\Gamma(\tau)}{\Gamma(0)}
\tag{5-23}
$$

Therefore, the intensity of the resulting electric field can be rewritten as

$$
\begin{aligned}
I(\tau,t) &= I_1 + I_2 + 2K_1 K_2 \Gamma(0)\,\mathrm{Re}[\gamma(\tau)] \\
&= I_1 + I_2 + 2\sqrt{I_1 I_2}\,\mathrm{Re}[\gamma(\tau)]
\end{aligned}
\tag{5-24}
$$

This equation can be further simplified by writing $\gamma(\tau)$ as a magnitude and a phase,

$$\gamma(\tau) = |\gamma(\tau)| e^{j\phi(\tau)}$$
$$= |\gamma(\tau)| e^{j[\Delta\phi(\tau) + \Delta\delta\phi_0(\tau)]}$$

, (5-25)

where $|\gamma(\tau)|$ is called the degree of temporal coherence, $\Delta\phi(\tau)$ is the phase difference related to the optical frequency and the delay time, and $\Delta\delta\phi_0(\tau)$ is the phase noise of the beat signal associated with the laser phase noise and the delay time. Finally, the intensity of the resulting electric field can be rewritten as

$$I(\tau, t) = I_1 + I_2 + 2\sqrt{I_1 I_2}|\gamma(\tau)|\cos[\Delta\phi(\tau) + \Delta\delta\phi_0(\tau)]$$

$$= (I_1 + I_2)\left\{1 + \frac{2\sqrt{I_1 I_2}}{I_1 + I_2}|\gamma(\tau)|\cos[\Delta\phi(\tau) + \Delta\delta\phi_0(\tau)]\right\}$$

$$= I_0\{1 + V'\cos[\Delta\phi(\tau) + \Delta\delta\phi_0(\tau)]\}$$

, (5-26)

where I_0 is the average intensity of the resulting electric field ($I_0 = I_1 + I_2$), and V' is the contrast of the beat signal given by

$$V' = \frac{2\sqrt{I_1 I_2}}{I_1 + I_2}|\gamma(\tau)|.$$

(5-27)

Comparing Equation (5-27) with Equation (2-20), we can see that the contrast of the beat signal produced by the real optical FMCW waves is modified by the degree of temporal coherence $|\gamma(\tau)|$, and the initial phase is modified by a quantity $\Delta\delta\phi_0(\tau)$. If the intensities of the two interfering beams are equal, the contrast simply equals the degree of temporal coherence

$$V' = |\gamma(\tau)|.$$

(5-28)

In general, the degree of temporal coherence takes on a value between 0 and 1. The optical source is coherent when $|\gamma(\tau)| = 1$ and completely incoherent when $|\gamma(\tau)| = 0$. The optical source is said to be partially coherent for other values. No beat signal is observed with an incoherent optical source, and the contrast is reduced with a partially coherent optical source.

It should be noted that the Fourier transform of the power spectrum of an optical source equals the self-correlation of the electric field. Hence, Equations (5-27) and (5-10) actually are equivalent. In addition, because the period used for calculating the intensity of an optical FMCW wave is much longer than the temporal period of the optical wave (T) but much shorter than the period of the beat signal (T_b), when two or more practical optical FMCW waves interfere, the phase of the beat signal $\Delta\phi(\tau)$ still is a function of time because the optical frequency varies with time, and the phase noise of the resulting field $\Delta\delta\phi_0(\tau)$ still can be a measurable random quantity, which will be discussed further in the next section. This might be the major difference between optical FMCW interference and optical homodyne interference.

5.3 Influence of the Phase Noise of Optical Sources

As indicated in Chapter 3, the optical FMCW wave emitted from any real optical source always contains noises in intensity and phase. In general, the intensity noise of an optical source (such as a semiconductor laser) is a comparatively slow variation, but the phase noise of an optical source is a rapidly changing random quantity that may affect the optical FMCW interference significantly.

When the phase noise is considered, the real phase $\phi_r(t)$ of a practical optical FMCW wave should be expressed by

$$\phi_r(t) = \int_0^t \omega(t)dt + \delta\phi_0(t)$$

$$= \phi(t) + \delta\phi_0(t) \qquad , \qquad (5\text{-}29)$$

where $\phi(t)$ is the regular phase component defined as before and $\delta\phi_0(t)$ is the phase noise of the optical source (a random quantity representing the fluctuation of the phase from the regular waveform). The wave function of the practical optical FMCW wave can be written as

$$E(\tau,t) = E_0 e^{j[\phi(t-\tau)+\delta\phi_0(t-\tau)]} \qquad , \qquad (5\text{-}30)$$

where E_0 is the amplitude and τ is the propagation time of the wave from the optical source to the point in space under consideration.

If two optical FMCW waves, derived from the same optical source but propagated along different paths, are recombined to interfere, the intensity of the resulting electric field will be

$$\begin{aligned}
I(\tau_1,\tau_2,t) &= |E_1(\tau_1,t) + E_2(\tau_2,t)|^2 \\
&= [E_1(\tau_1,t) + E_2(\tau_2,t)][E_1(\tau_1,t) + E_2(\tau_2,t)]* \\
&= E_1(\tau_1,t)E_1*(\tau_1,t) + E_2(\tau_2,t)E_2*(\tau_2,t) \\
&\quad + E_1(\tau_1,t)E_2*(\tau_2,t) + E_1*(\tau_1,t)E_2(\tau_2,t)
\end{aligned}$$

$$
\begin{aligned}
&= I_1 + I_2 + 2\sqrt{I_1 I_2}\, \cos[\phi(t - \tau_1) - \phi(t - \tau_2) \\
&\quad + \delta\phi_0(t - \tau_1) - \delta\phi_0(t - \tau_2)] \\
&= I_0\{1 + V\cos[\phi(t - \tau_1) - \phi(t - \tau_2) + \delta\phi_0(t - \tau_1) - \delta\phi_0(t - \tau_2)]\}
\end{aligned}
$$

$$(5\text{-}31)$$

where $E_1{}^*(\tau_1, t)$ and $E_2{}^*(\tau_2, t)$ are the conjugates of $E_1(\tau_1, t)$ and $E_2(\tau_2, t)$, respectively; I_1, E_{01}, and τ_1 are the intensity, amplitude, and propagation time of the first wave, respectively ($I_1 = E_1(\tau_1, t)E_1{}^*(\tau_1, t) = E_{01}{}^2$); and I_2, E_{02}, and τ_2 are the intensity, amplitude, and propagation time of the second wave, respectively ($I_2 = E_2(\tau_2, t)E_2{}^*(\tau_2, t) = E_{02}{}^2$). I_0 is the average intensity of the resulting electric field ($I_0 = I_1 + I_2$), and V is the contrast of the beat signal ($V = 2\sqrt{I_1 I_2}/(I_1 + I_2)$).

Performing a variable transformation $t = t - \tau_1$, the above equation can be rewritten as

$$
\begin{aligned}
I(\tau, t) &= I_0\{1 + V\cos[\phi(t) - \phi(t - \tau) + \delta\phi_0(t) - \delta\phi_0(t - \tau)]\} \\
&= I_0\{1 + V\cos[\Delta\phi(\tau, t) + \Delta\delta\phi_0(\tau, t)]\}
\end{aligned}
$$

$$(5\text{-}32)$$

where τ is the delay time of the second wave with respect to the first one ($\tau = \tau_2 - \tau_1$), $\Delta\phi(\tau, t)$ is the regular phase of the beat signal given by

$$\Delta\phi(\tau, t) = \phi(t) - \phi(t - \tau)$$

$$(5\text{-}33)$$

and $\Delta\delta\phi_0(\tau, t)$ is the phase noise of the beat signal given by

$$\Delta\delta\phi_0(\tau, t) = \delta\phi_0(t) - \delta\phi_0(t - \tau)$$

$$(5\text{-}34)$$

Comparing Equation (5-32) with Equation (2-43), it can be seen that the phase noise of the real optical source causes a phase noise in the beat signal. The experimental results show that the phase noise of the beat signal from an optical FMCW interferometric system generally is still a measurable random quantity, which, of course, will reduce the measurement accuracy significantly.

In general, the statistical property of a noise is represented by its standard deviation. Assuming the laser noise is an ergodic stationary random process, based on Equation (5-34), the standard deviation σ_b of the phase noise of the beat signal can be written as

$$\sigma_b^2 = 2\sigma_l^2$$

$$(5\text{-}35)$$

where σ_l is the standard deviation of the phase noise of the laser.

Note that it is extremely easy to enlarge laser noise via feedback light from the optical system. Moreover, component vibration and scattering light will introduce additional phase noise to the beat signal. Therefore, the phase noise of the beat signal usually increases with the optical path difference.

Chapter 6

Optical Frequency-Modulated Continuous-Wave Interferometers

Optical interferometers are instruments or devices in which two or more optical waves derived from the same optical source but traveling along separate paths are combined to produce an interferometric signal. Optical interferometers play an important role in the field of precision measurements. Based on the interference phenomenon of optical waves, many physical quantities, such as displacement, optical wavelength, and optical reflective index, can be precisely determined.

This chapter will first discuss the requirements, procedures, and special considerations for constructing optical FMCW interferometers and then introduce some double-beam amplitude-division FMCW interferometers, including the Michelson FMCW interferometer, the Mach-Zehnder FMCW interferometer, and the Fabry-Perot FMCW interferometer.

6.1 Construction of Optical FMCW Interferometers

The essential requirement for constructing an interferometer is that there should be an optical arrangement in which two or more beams, derived from the same optical source but traversing separate paths, can be recombined to interfere. According to the number of the optical waves involved in the interference, interferometers can be categorized into *double-beam interferometers* and *multiple-beam interferometers*.

In addition, in practice, there can be two approaches to splitting and recombining the interfering beams: *wave-front division* and *amplitude division*. In wave-front division interference, two or more parts of a wave front are selected and then redirected to a common volume in space to interfere, while in amplitude division interference, the original wave-front amplitude is split into two or more parts, each part is directed along a different path, and then they are recombined to interfere. Consequently, interferometers can also be classified as *wave-front division interferometers* and *amplitude division interferometers*.

According to the analysis in Chapter 2, it can be easily seen that, for constructing optical FMCW interferometers, the following additional requirements should also be satisfied:

(1) The interfering beams should be collimated and should be parallel to each other when they are interfering, so that the spatial dependence of the intensity of the resulting electric field can be removed and a beat signal with an adequate contract can be obtained.

(2) The interferometer should be unbalanced, so that the beat signal with a proper frequency can be obtained. In the case of sawtooth-wave FMCW interference, for instance, if the frequency of the beat signal is desired to be equal to or larger than the frequency of the modulation waveform ($v_b \geq v_m$) (i.e., the beat signal has at least one full wavelet in each modulation period), according to Equation (2-64), the optical path difference between the two interfering beams, *OPD*, should be

$$OPD \geq \frac{c}{\Delta v},$$

(6-1)

where *c* is the speed of light in free space, and Δv is the optical frequency modulation excursion.

(3) The interferometer should include a photodetector for converting the optical signal into an electric signal in order that the time-domain interference pattern can be viewed by using an electronic oscilloscope or processed by using an electric circuit or computer.

Optical FMCW interferometers have their specific characteristics, and therefore to construct an optical FMCW interferometer requires a procedure like this:

(1) Construct a collimated-beam unbalanced homodyne interferometer with a laser without modulating its frequency, as shown in Figure 6.1(a).

(2) Adjust the optical elements until only one bright fringe is left in the field of view, and place a photodetector at the center of this bright fringe, as shown in Figure 6.1(b).

(3) Modulate the frequency of the laser with a proper modulation waveform, and use an electronic oscilloscope to view the pattern of the electric beat signal, as shown in Figure 6.1(c).

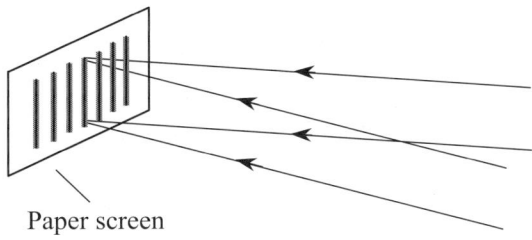

(a) Construct a collimated-beam unbalanced homodyne interferometer.

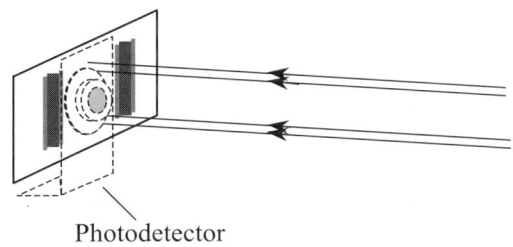

(b) Adjust to one bright fringe and place a photodetector at the center.

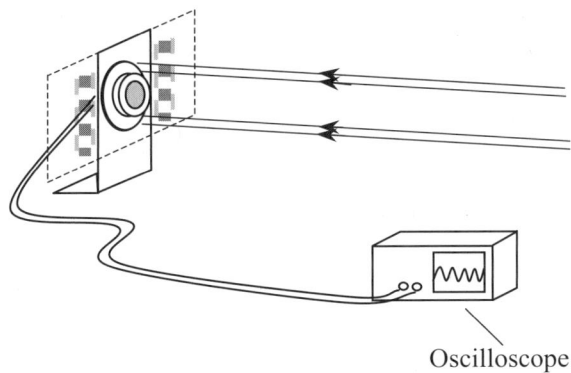

(c) Modulate the laser frequency and view the pattern of the electric beat signal with an oscilloscope.

Figure 6.1. Construction of optical FMCW interferometers.

When constructing an optical FMCW interferometer with a semiconductor laser, we must beware the feedback light from the interferometer. As indicated in Chapter 3, the semiconductor laser is very easily disturbed by the feedback light. A little portion of the light reflected from the uncoated surfaces of the coupling lens is enough to cause a high level of noise. In practice, slight misalignment of the optical axes of the components can reduce the feedback light, but this may cause difficulty in collimating the output beam. Moreover, this method is useless for a reflectometric optical system. Hence, the best way to prevent feedback light for transmissive optical systems is to coat the coupling lenses with an antireflection film; for the reflectometric optical systems, an additional optical isolator should be used.

The optical isolator is a magneto-optic device that is based on the Faraday effect. The polarization direction of a linearly polarized optical wave incident on a piece of glass will be rotated if a strong magnetic field is applied on the glass in the propagation direction. The rotation angle θ can be determined by the empirical law

$$\theta = VBl, \tag{6-2}$$

where V is a material-dependent constant (called the Verdet constant), B is the static magnetic field, and l is the length of the glass, as shown in Figure 6.2. The interesting thing is that the rotation direction is dependent on the direction of the magnetic field, but not on the propagation direction of the incident light.

The typical optical isolator is made up of two polarizers (P_1 and P_2) and a $45°$ Faraday rotator (R) situated in the middle, as shown in Figure 6.3. The transmission axis of P_2 is rotated $45°$ with respect to the transmission axis of P_1. For the forward transmission, the incident optical wave is first polarized by P_1 and its polarization direction is then rotated clockwise by $45°$ from R. Since the polarization direction of the emerging light from R is the same as the transmission axis of P_2, the optical wave can pass the isolator, as shown in Figure 6.3(a). For the backward transmission, the incident optical wave undergoes the same process, but the polarization direction of the emerging light from R after $45°$ clockwise rotation is perpendicular to the transmission axis of P_1, and thus it cannot be propagated, as shown in Figure 6.3(b). Because the

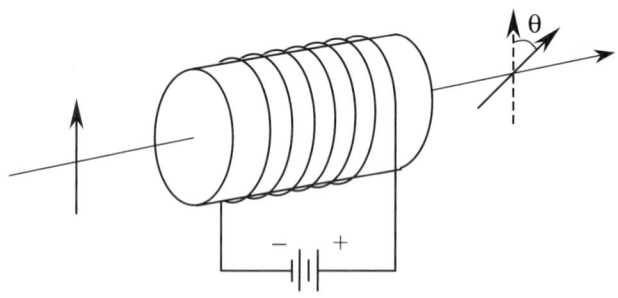

Figure 6.2. The Faraday effect.

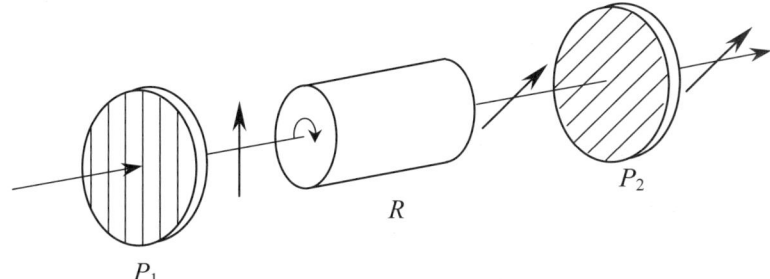

(a) Transmission of a forward propagating optical wave.

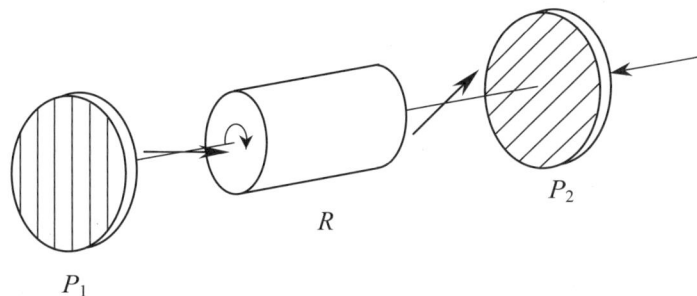

(b) Isolation of a backward propagating optical wave.

Figure 6.3. Principle of the optical isolator.

optical wave passing through the isolator is incidentally polarized, this type of the optical isolator is also called a polarization-dependent optical isolator.

The performance of the optical isolator generally is evaluated by its insertion loss (the reciprocal of the forward transmittance in decibels) and extinction ratios (the ratio of the backward transmittance and the forward transmittance in decibels). The typical values for commercial polarization- dependent optical isolators are 0.2 dB in insertion loss and −40 dB in extinction ratio.

In the following sections, we will introduce some double-beam amplitude-division FMCW interferometers, which are derived from the classical interferometers such as the Michelson interferometer, the Mach-Zehnder interferometer, and the Fabry-Perot interferometer. (For convenience, we call them the Michelson FMCW interferometer, the Mach-Zehnder FMCW interferometer, and the Fabry-Perot FMCW interferometer, respectively.) Wave-front division interferometers, however, are rarely used in optical FMCW interference because they are not convenient for constructing collimated-beam unbalanced interferometers. Multiple-beam interferometers are also seldom used in optical

FMCW interference due to their problems of mutual interference and signal complexity.

6.2 The Michelson FMCW Interferometer

The Michelson FMCW interferometer (or called the Twyman-Green FMCW interferometer) is shown schematically in Figure 6.4. An FMCW laser beam is first collimated by a collimating lens and then is divided into two beams by a beam splitter (*BS*). The reference beam propagates along the path l_1 and is reflected by a fixed mirror (M_1), while the signal beam propagates along the path l_2 and is reflected by a moving mirror (M_2). These two reflected beams are recombined by the same beam splitter *BS* to produce a beat signal, and the beat signal is received by a photodetector.

The optical path difference *OPD* between the two interfering beams can be written as

$$OPD = 2n(l_2 - l_1),$$

(6-3)

where n is the optical refractive index of air ($n \approx 1$), and l_1 and l_2 are distances from *BS* to M_1 and M_2, respectively.

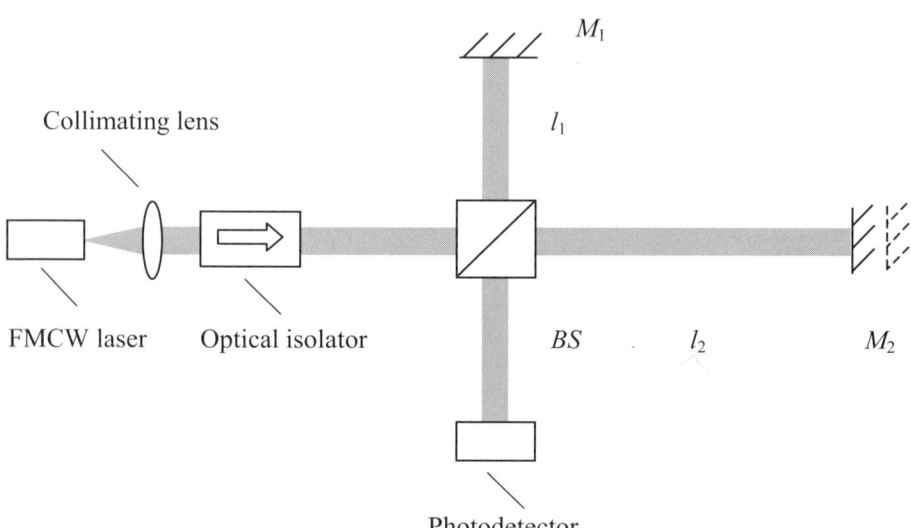

Figure 6.4. The Michelson FMCW interferometer.

The Michelson FMCW interferometer usually is used for measuring the displacement of an object. For instance, if the frequency of the laser is modulated with a sawtooth waveform, the intensity $I(t)$ of the beat signal detected in any modulation period can be written as

$$I(OPD,t) = I_0 \left[1 + V \cos\left(\frac{2\pi \Delta v v_m OPD}{c} t + \frac{2\pi}{\lambda_0} OPD \right) \right]$$

$$= I_0 [1 + V \cos(2\pi v_b t + \phi_{b0})] \qquad , \qquad (2\text{-}63)$$

where I_0 is the average intensity of the beat signal, V is the contrast of the beat signal, Δv is the optical frequency modulation excursion, v_m is the modulation frequency, OPD is the optical path difference, c is the speed of light in free space, λ_0 is the central optical wavelength, and v_b and ϕ_{b0} are the frequency and initial phase of the beat signal, respectively.

Considering Equation (6-3), the frequency of the beat signal v_b can be written as

$$v_b = \frac{2n(l_2 - l_1)\Delta v v_m}{c} \qquad , \qquad (6\text{-}4)$$

and the initial phase of the beat signal ϕ_{b0} can be written as

$$\phi_{b0} = \frac{4\pi n(l_2 - l_1)}{\lambda_0} . \qquad (6\text{-}5)$$

Hence, measuring the frequency of the beat signal v_b, we can obtain the absolute value of the path difference of the two mirrors $(l_2 - l_1)$,

$$l_2 - l_1 = \frac{c}{2n\Delta v v_m} v_b \qquad , \qquad (6\text{-}6)$$

and measuring the phase shift of the beat signal $\Delta\phi_{b0}$, we can obtain the displacement of the moving mirror Δl_2

$$\Delta l_2 = \frac{\lambda_0}{4\pi n} \Delta\phi_{b0} . \qquad (6\text{-}7)$$

Figure 6.5 shows another version of the Michelson FMCW interferometer, which employs a big beam splitter (BS) to split and recombine the light beams and two large retroreflectors (R_1 and R_2) to reflect the reference and signal beams,

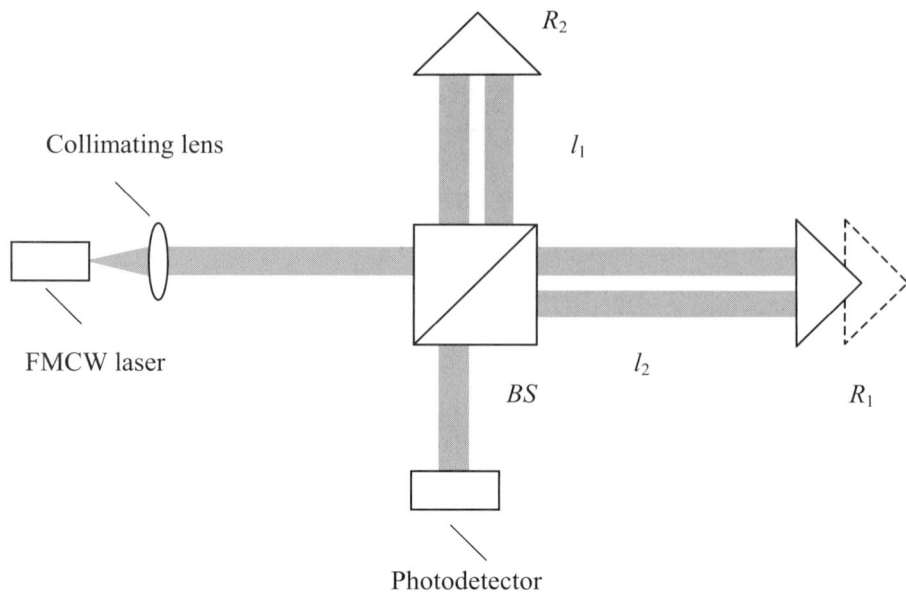

Figure 6.5. The Michelson FMCW interferometer with retroreflectors.

respectively, back to the system. One of the advantages of this configuration is that the reflected beams are immune to the influence of the direction and position of the retroreflectors because the retroreflectors always reflect the beams by 180°. Another advantage of this configuration is that there is no feedback light to affect the laser because the forward-propagating beams and the reflected beams travel along different paths. Therefore, for this configuration, the optical isolator can be omitted.

6.3 The Mach-Zehnder FMCW Interferometer

The Mach-Zehnder FMCW interferometer, as shown in Figure 6.6, employs two beam splitters (BS_1 and BS_2) and two mirrors (M_1 and M_2) to divide and recombine the beams. A collimated FMCW laser beam is first divided into two beams by BS_1. One is reflected by M_1 and passes through BS_2, propagating toward the photodetector; another is reflected by M_2 and BS_2 consecutively. These two beams mix coherently behind BS_2, and the beat signal produced is received by a photodetector

The optical path difference OPD between the two interfering beams can be written as

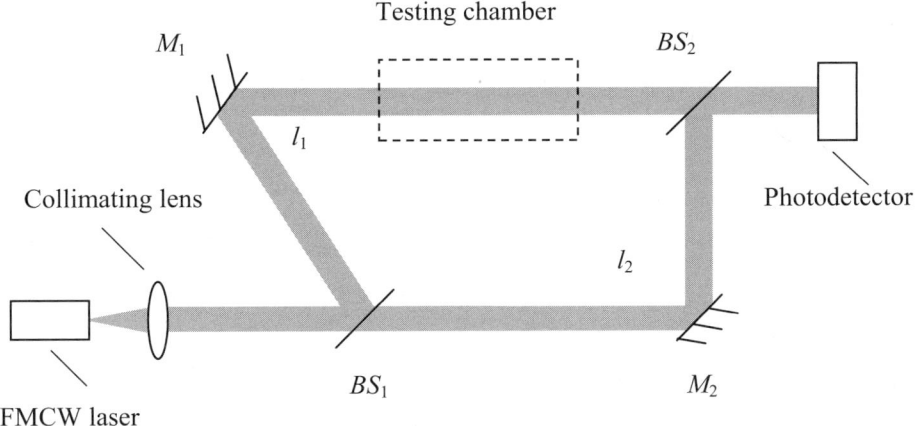

Figure 6.6. The Mach-Zehnder FMCW interferometer.

$$OPD = n(l_2 - l_1),$$ (6-8)

where n is the optical refractive index of air ($n \approx 1$), l_1 is the total geometrical path length from BS_1 to M_1 and BS_2, and l_2 is the total geometrical path length from BS_1 to M_2 and BS_2, assuming the beam splitters BS_1 and BS_2 are identical. The signal analysis for the Mach-Zehnder FMCW interferometer is similar to that for the Michelson FMCW interferometer, and therefore we will not repeat it here.

The advantage of the Mach-Zehnder FMCW interferometer is that all the laser beams propagate forward so that there is no feedback light to influence the radiation property of the laser. The disadvantage of the Mach-Zehnder FMCW interferometer is that the mirrors are not easy to remove without disturbing the parallel-beam interference arrangement. Therefore, the Mach-Zehnder FMCW interferometer is usually used for measuring the optical refractive index of a gas or liquid in a transparent chamber that is placed in a path of the interferometer.

6.4 The Fabry-Perot FMCW Interferometer

The Fabry-Perot FMCW interferometer (or called the Fizeau FMCW interferometer) is shown schematically in Figure 6.7. A collimated FMCW laser beam passes through a beam splitter (*BS*) and is reflected by a partially reflecting mirror (*PM*). The remainder of the laser beam propagates through a length of air path *l* and then is reflected by a mirror (*M*). The two reflected laser beams propagate back to the interferometer to mix coherently. A portion of the beat signal produced is reflected by *BS* and finally detected by a photodetector.

The optical path difference *OPD* between the two reflected beams can be written as

$$OPD = 2nl$$
,
(6-9)

where *n* is the optical refractive index of air ($n \approx 1$) and *l* is the distance from *PM* to *M*.

The Fabry-Perot FMCW interferometer is suitable for measuring the distance or displacement of an object. In addition, because of the nature of the single-arm interferometer, the Fabry-Perot FMCW interferometer is more compact than the Michelson FMCW interferometer.

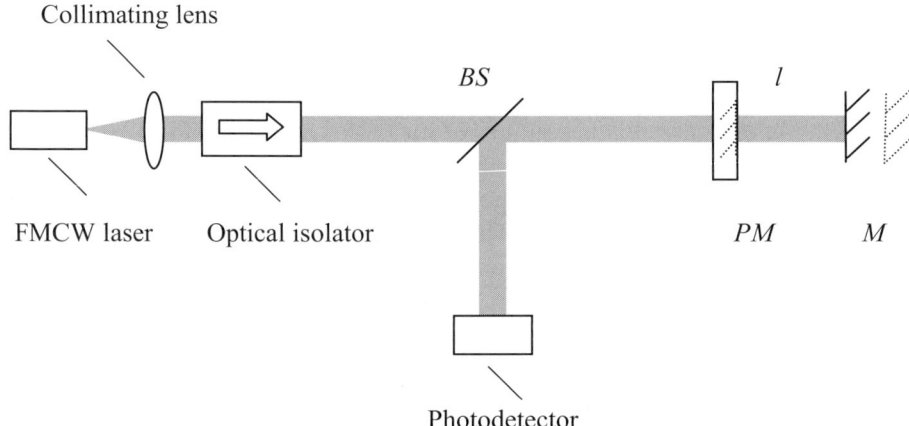

Figure 6.7. The Fabry-Perot FMCW interferometer.

Chapter 7

Fiber-optic Frequency-Modulated Continuous-Wave Interferometers

Optical fibers have been widely used in the fields of image transmission and optical communication, providing a low-noise, low-attenuation, low-cost, long-distance, and flexible light-propagation medium. Application of optical fibers and fiber-optic components to optical interferometers makes the interferometers compact, reliable, flexible, and more accurate. In addition, with optical fiber technology, more advanced detection techniques, more sophisticated interferometers, and even "solid" interferometers (i.e., all-fiber interferometers) can be developed.

In this chapter, we will first briefly discuss optical fibers and fiber-optic components and then introduce some fiber-optic FMCW interferometers, including fiber-optic Michelson FMCW interferometers, fiber-optic Mach-Zehnder FMCW interferometers, and fiber-optic Fabry-Perot FMCW interferometers.

7.1 Introduction to Optical Fibers

Optical fibers are cylindrical dielectric optical waveguides that generally consist of a core of highly refractive transparent material, a cladding of low- refraction transparent material, and a plastic protective jacket, as shown in Figure 7.1. For some specific applications, optical fibers may deviate slightly from this symmetry.

Optical fibers can be categorized into two basic types, step-index fibers and gradient-index fibers. In a step-index fiber, the propagation of light is based on the total internal reflection at the core-cladding interface. According to Snell's law, the critical acceptance angle θ_c for incident light is equal to

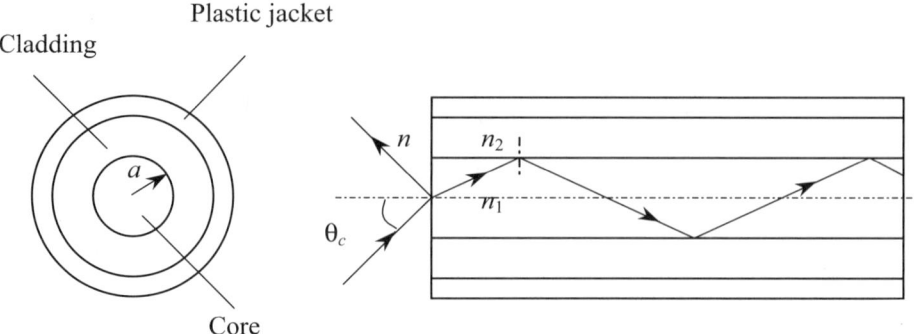

(a) Cross section of an optical fiber.   (b) Propagation of light in an optical fiber.

Figure 7.1. Profile of an optical fiber.

$$\theta_c = \arcsin\left[\frac{\sqrt{n_1^2 - n_2^2}}{n}\right],$$

$$(7\text{-}1)$$

where n_1 is the refractive index of the core, n_2 is the refractive index of the cladding, and n is the refractive index of air. Usually, we use the numerical aperture NA to describe the critical acceptance angle

$$NA = n \sin \theta_c$$
$$= \sqrt{n_1^2 - n_2^2}.$$

$$(7\text{-}2)$$

Moreover, the light propagating in an optical fiber must satisfy a self-interference condition in order to be trapped in the waveguide. Therefore, it can be imagined that only a finite number of paths in an optical fiber satisfy this condition. These paths are usually called *modes*.

Geometrical optics only gives an approximate description of optical fibers. The actual propagation characteristics of optical fibers (particularly for the fibers that support a small number of modes) must be understood in the context of guided-wave optics (i.e., by solving Maxwell's equations with the appropriate boundary conditions of the fiber structure).

For optical fibers, which exhibit a small change in optical refractive index at the boundary between the core and the cladding, the electric field can be described by a scalar wave equation

$$\nabla^2 E = \frac{n(r)^2}{c^2} \frac{\partial^2 E}{\partial t^2}$$

(7-3)

where E is the electric field, $n(r)$ is the optical refractive-index distribution, and c is the speed of light in free space. In the cylindrical coordinate system of the fiber, the electric field $E(r, \theta, z)$ is assumed to be of the form

$$E(r,\theta,z) = R(r)e^{\pm jm\phi}e^{j(\omega t - \beta z)}$$

(7-4)

where m is the azimuthal mode number ($m = 0, 1, 2, \ldots$), ω is the angular frequency of the optical wave, β is the propagation number of the beam, and $R(r)$ is the radial part of the scalar field that satisfies the eigen-value equation

$$r^2 \frac{d^2 R}{dr^2} + r \frac{dR}{dr} + \{[k_0^2 n^2(r) - \beta^2]r^2 - m^2\}R = 0$$

(7-5)

where k_0 is the propagation number in free space.

In the case of the step-index fiber, the solution of Equation (7-5) in the core propagation region ($r < a$) is the Bessel function

$$R(r) = A J_m\left(\frac{ur}{a}\right)$$

(7-6)

while A is a constant, J_m is the mth Bessel function, a is the radius of the core, and u is a parameter of the fiber; the solution in the cladding ($r > a$) is the modified Bessel function

$$R(r) = B K_m\left(\frac{wr}{a}\right)$$

(7-7)

where B is a constant, K_m is the mth modified Bessel function, and w is another parameter of the fiber. The fiber parameters u and w are defined as

$$\left(\frac{u}{a}\right)^2 = n_1^2 k_0^2 - \beta^2$$

(7-8)

$$\left(\frac{w}{a}\right)^2 = \beta^2 - n_2^2 k_0^2$$

(7-9)

where $n_2 k_0 \leq \beta \leq n_1 k_0$, and u and w are related to the normalized frequency V by

$$
\begin{aligned}
V^2 &= u^2 - w^2 \\
&= a^2 k_0^2 (n_1^2 - n_2^2) \\
&= 2a^2 k_0^2 n_1^2 \Delta
\end{aligned}
\tag{7-10}
$$

where Δ is the relative refractive index difference given by

$$
\begin{aligned}
\Delta &= \frac{n_1^2 - n_2^2}{2n_1^2} \\
&\approx \frac{n_1 - n_2}{n_1}
\end{aligned}
\tag{7-11}
$$

Considering the continuity condition at the core-cladding boundary and decay requirement outside the core region for guided modes, we have the characteristic equation

$$
\frac{u J_{m-1}(u)}{J_m(u)} = -\frac{w K_{m-1}(w)}{K_m(w)}
\tag{7-12}
$$

Solving these equations, we can obtain the propagation number β as well as the parameters u, w, and A. In addition, using these equations, we can also ascertain the modes in an optical fiber.

According to the property of the modified Bessel function, if $w < 0$, $K_m(wr/a)$ enlarges as r increases, and thus it represents an optical wave propagating in the cladding (or *radiation mode*), while if $w > 0$, $K_m(wr/a)$ decreases as r increases, and thus it represents an optical wave propagating in the core (or *guided mode*). Obviously, the cutoff of the mode should be $w = 0$, and $w \to \infty$ stands for the optical wave far from the cutoff. Therefore, the possible guided linear polarization modes (or LP modes) can be differentiated by any two adjacent roots of the equations $w = 0$ and $w = \infty$. Considering Equation (7-12), these equations are equivalent to the equations $J_{m-1}(u) = 0$ and $J_m(u) = 0$. Hence, the LP modes in a step fiber are actually determined by any two adjacent roots of the equations $J_{m-1}(u) = 0$ and $J_m(u) = 0$.

Figure 7.2 illustrates the first few zeroes of $J_0(u)$ and $J_1(u)$, as well as the corresponding linear polarization guided modes. Figure 7.3 shows the intensity distribution patterns of some low-order modes in step-index optical fibers. Obviously, only the LP_{01} mode has a uniform phase front and no zero intensity across the field.

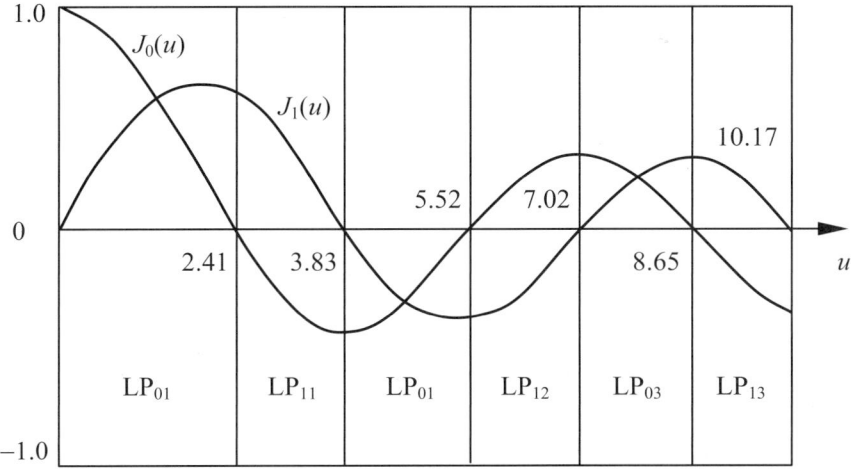

Figure 7.2. The roots of the zero- and first-order Bessel functions.

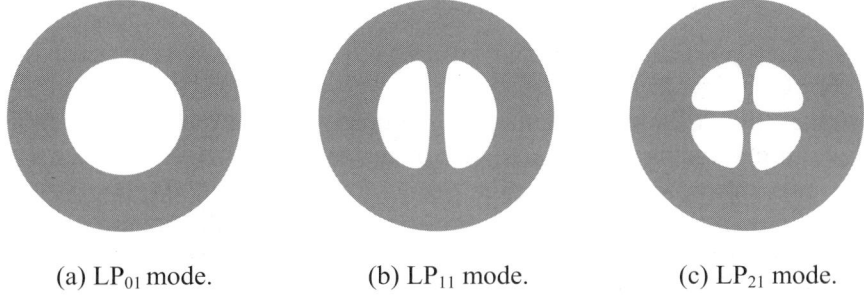

(a) LP_{01} mode.   (b) LP_{11} mode.   (c) LP_{21} mode.

Figure 7.3. Intensity distribution patterns of some low-order modes in step-index optical fibers.

7.1.1 Single-Mode Optical Fibers

Single-mode optical fibers are optical fibers that support only one fundamental mode. If the normalized frequency V of a step-index fiber is smaller than 2.405, only one mode (the LP_{01} mode) can exist in the fiber. Therefore, the single-mode propagation condition for a step-index fiber is

$$V < 2.45 .$$

(7-13)

Considering Equation (7-10), the maximum core radius for single-mode propagation should be

$$a = \frac{2.405\lambda}{2\pi\sqrt{n_1^2 - n_2^2}}$$
$$\approx \frac{2.405\lambda}{2\pi n_1 \sqrt{2\Delta}},$$

(7-14)

For instance, assuming $\lambda = 0.660$ μm, $n_1 = 1.45$, and $\Delta = 0.003$, we find that the maximum core radius for single-mode operation is 2.25 μm. In fact, the core diameter of the real single-mode optical fiber is around 5 μm.

For convenience, we frequently use the effective refractive index to describe the propagation speed of the light in a guided mode. For instance, the effective refractive index n_e of the LP_{01} mode can be defined as

$$n_e = \frac{\beta}{k_0},$$

(7-15)

where β is the propagation number of the LP_{01} mode and k_0 is the propagation number of the light in free space.

In fact, each LP mode contains several precise vector modes that can be derived from the vector wave equation. For instance, the LP_{01} mode actually includes two orthogonal polarization modes, the HE_{11}^{x} mode and the HE_{11}^{y} mode, as shown in Figure 7.4, where the arrow indicates the polarization direction of the guided wave (i.e., the direction of the electric field vector). In other words, a single-mode fiber actually enables two orthogonal polarization waves to propagate.

In the practical situation, there is always a small birefringence in the single-mode fiber due to imperfections of cylindrical symmetry or internal stress. This internal birefringence generally is irregularly distributed and easily magnified by environmental conditions, such as tension, twisting or bending. Therefore, the polarization state of the light beam in the single-mode fiber may vary unpredictably.

7.1.2 Birefringent Optical Fibers

Birefringent fibers (also called polarization-maintaining fibers) are single-mode optical fibers that have a large regular internal birefringence and can maintain the polarization state of the light propagating in them. The reason is that the two orthogonal polarization components travel along the birefringent fiber with significantly different velocities and thus prevent the transfer of light

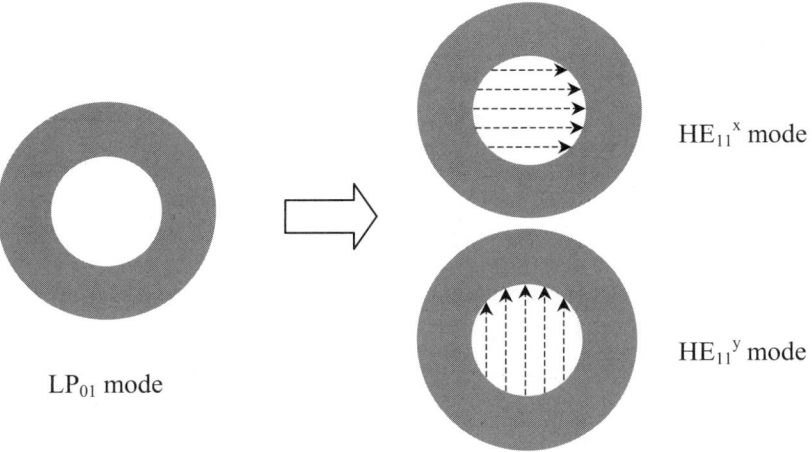

Figure 7.4. Polarization modes in a single-mode fiber.

energy from one mode into the other.

If a linearly polarized beam is coupled to one of its orthogonal polarization modes, the state of polarization is maintained, as shown in Figure 7.5(a). On the other hand, if a linearly polarized beam enters the birefringent fiber at an angle with the principal axes (i.e., the polarization directions of the two orthogonal polarization modes), the polarization state of the light will periodically change from a linear state to an elliptical state, a circular state, and back to the initial linear state over a length that is defined as the beat length of the birefringent fiber in which the two orthogonal polarization modes differ in phase by 2π, as shown in Figure 7.5(b). The beat length Λ can be determined by

$$\Lambda = \frac{2\pi}{\left| \beta_x - \beta_y \right|},$$

$$(7\text{-}16)$$

where β_x and β_y are the propagation numbers of the $HE_{11}{}^x$ mode and the $HE_{11}{}^y$ mode, respectively.

There can be two approaches to introducing birefringence in the core of optical fiber: One involves prestressing the core, either by means of an elliptical cladding or by fabricating the fiber with two regions of highly doped glass located on opposite sides of the core, as in panda-type fiber or "bow tie" fiber, as shown in Figures 7.6(a) to 7.6(c); the other involves fabricating the core with an asymmetry as in elliptical-core fiber or double-core fiber, shown in Figures 7.6(d) and 7.6(e). The elliptical-core circular-cladding fiber, as shown

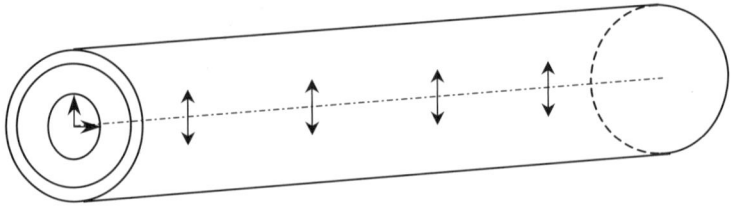

(a) Propagation along one principal axis.

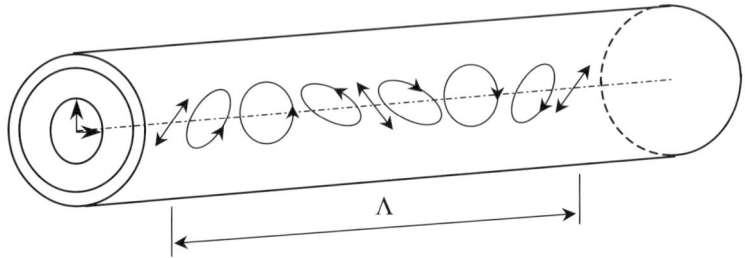

(b) Propagation along two principal axes.

Figure 7.5. Propagation of light in a birefringent fiber.

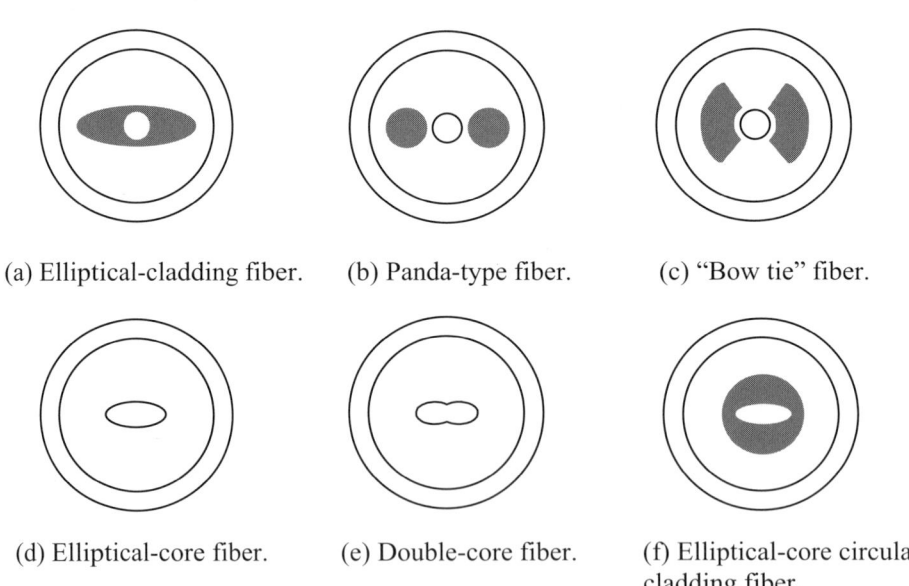

(a) Elliptical-cladding fiber.   (b) Panda-type fiber.   (c) "Bow tie" fiber.

(d) Elliptical-core fiber.   (e) Double-core fiber.   (f) Elliptical-core circular-cladding fiber.

Figure 7.6. Profiles of birefringent fibers.

in Figure 7.6(f), has both stress-induced birefringence and asymmetry-induced birefringence.

Similarly, for the sake of convenience, we use n_{ex} and n_{ey} to represent the effective refractive index of the HE_{11}^{x} mode and the HE_{11}^{y} mode, respectively,

$$n_{ex} = \frac{\beta_x}{k_0},$$

(7-17)

$$n_{ey} = \frac{\beta_y}{k_0},$$

(7-18)

where k_0 is the propagation number of the light in free space. Moreover, we also frequently use n_{ef} to represent the effective refractive index of the fast mode (i.e., the mode with a smaller effective refractive index or smaller propagation number) and use n_{es} to represent the effective refractive index of the slow mode (i.e., the mode with a bigger effective refractive index or bigger propagation number) in a birefringent fiber.

A single-mode optical fiber, like an optical spatial filter, can regulate the spatial intensity distribution of the incident beam. The light propagating in a single-mode fiber, like a plane wave, can propagate for a long distance with very little distortion. In other words, no matter what the incident beam is and no matter how long the optical fiber, once the light is coupled into a single-mode fiber, the spatial distribution of the intensity of the emerging beam is almost the same. Since single-mode optical fibers have such superior properties, they are widely used in various fiber-optic interferometers.

The transmission characteristics of optical fibers include dispersion and attenuation. Single-mode fibers have material dispersion and intramodal dispersion, but no intermodal dispersion. Material dispersion is an intrinsic property of the glass used in the core and cladding. Intramodal dispersion comprises waveguide dispersion and profile dispersion. Waveguide dispersion is due to the fact that a change in frequency will change the propagation number for a particular mode, while profile dispersion is due to the fact that a change in frequency will change the index profile of the core and cladding and therefore cause a small but measurable dispersion. In general, profile dispersion is smaller than waveguide dispersion or material dispersion, and waveguide dispersion has the sign opposite to material dispersion. Therefore, material, waveguide, and profile dispersions act together, so that the total dispersion may vanish at a wavelength.

There are two major factors that cause light attenuation in optical fibers: material absorption and scatting loss. Material absorption includes metallic ion impurity absorption, OH ion absorption, and long-wavelength molecular vibration absorption. Scattering loss is mainly caused by Rayleigh scattering, which is in almost all directions and produces an energy loss proportional to $1/\lambda^4$. Over the past two or three decades, optical fiber fabrication technology has experienced a number of tremendous changes. Now, the metallic ion impurity and the OH ion dissolved in the glass can be reduced to an extremely low level. The total attenuation of the commercial optical fibers can be less than 0.2 dB/km.

For fiber-optic FMCW interferometers, dispersion in the optical fiber may cause a distortion of frequency modulation and both absorption and scattering in the optical fiber will cause attenuation of light energy, but scattering in the optical fiber will also introduce noise. Fortunately, the length of the optical fiber used in most fiber-optic FMCW interferometers is less than 100 meters, so that the effects of dispersion, absorption, and scattering usually are insignificant. However, for some specific fiber-optic interferometers, such as fiber-optic Sagnac gyroscopes, because the fiber length normally exceeds 1000 meters, the dispersion, absorption, and scattering in the optical fiber must always be considered.

Birefringent fibers generally have two uses in fiber-optic interferometers. The first use is to construct the truly single-mode fiber interferometers, in which only one polarized mode in the birefringent fiber is employed to propagate light. Because birefringent fibers can maintain the polarization state of light propagating in it and prevent energy coupling between the two orthogonal polarization modes, the truly single-mode fiber interferometers can avoid variation of the polarization state, resulting in better accuracy. The second use is to build the truly two-mode fiber interferometers, in which both orthogonal polarization modes in the birefringent fiber are employed to propagate light beams.

For convenience, in this book, we no longer distinguish between single-mode fiber interferometers as used here and true single-mode fiber interferometers because they are exactly the same in principle and configuration. However, if we mention birefringent fiber interferometers, we mean that they are true two-mode fiber interferometers, in which the two orthogonal polarization modes in the birefringent fiber will be employed to propagate light beams.

7.2 Introduction to Fiber-optic Components

Fiber-optic communications and fiber-optic sensors demand fiber-optic components capable of performing various functions, such as splitting, combining, isolating, polarizing, and so on. Usually, these functions are performed by taking the light out of the fiber, processing it by using bulk optical components, and then coupling the light back into the fiber. This would interrupt the light propagation in the fiber and lead to more energy loss and poor stability. With fiber-optic components, however, such problems can be properly solved because the processes can be performed inside the optical fiber without taking the light out of the optical fiber.

Thus far, various fiber-optic elements have been available. Note that some fiber-optic components are truly made of optical fibers (such as fiber-optic directional couplers), some fiber-optic components actually are precise mechanical components (such as fiber-optic connectors), and some fiber-optic components are just miniatures of bulk optical elements packed with optical fiber pigtails (such as fiber-optic magnetic optical isolators). In the following subsections, we will discuss the frequently used single-mode fiber-optic components, including fiber-optic directional couplers, fiber-optic polarizers, fiber-optic polarization controllers, fiber-optic connectors, and fiber-optic splices.

7.2.1 Fiber-optic Directional Couplers

Fiber-optic directional couplers (also called fiber-optic couplers or fiber couplers), like bulk optical beam splitters or combiners, can divide or recombine light beams. Fiber-optic couplers are based on the principle of waveguide coupling: When two fiber cores are brought sufficiently close to each other laterally, the modes of the two fibers will become coupled, and light power can transfer between the two fibers. The transferred power generally varies periodically with the product of the coupling length and the coupling coefficient (a measure of the strength of interaction between the two fibers, which depends on the fiber parameter, the separation of the cores, and the optical wavelength). If the propagation numbers of the modes of the individual fibers are equal, this power exchange can be complete. If their propagation numbers are different, there is still a periodic, but incomplete, exchange of power between the two fibers.

There are a number of parameters frequently used for describing the performance of a fiber-optic coupler:

(1) Coupling ratio (R): The ratio between the coupled power and the total output power as a percentage, or the ratio between the total output power and the coupled power in decibels,

$$R(\%) = \frac{P_c}{P_t + P_c} \times 100 \quad , \tag{7-19}$$

$$R(\text{dB}) = 10 \log\left(\frac{P_t + P_c}{P_c}\right), \tag{7-20}$$

where P_c is the coupled power and P_t is the transmitted power, as shown in Figure 7.7.

(2) Excess loss (L_e): The ratio between the input power and the total output power in decibels,

$$L_e(\text{dB}) = 10\log\left(\frac{P_i}{P_t + P_c}\right),$$

(7-21)

where P_i is the input power.

(3) Insertion loss (L_i): The ratio between the input power and the coupled power in decibels,

$$L_i(\text{dB}) = 10\log\left(\frac{P_i}{P_c}\right).$$

(7-22)

Obviously, L_i, L_e, and R are related by

$$L_i(\text{dB}) = R(\text{dB}) + L_e(\text{dB}).$$

(7-23)

(4) Directivity (D): The ratio between the back-coupled power and the input power in decibels,

$$D(\text{dB}) = 10\log\left(\frac{P_r}{P_i}\right),$$

(7-24)

where P_r is the power coupled back to the input end of the second fiber.

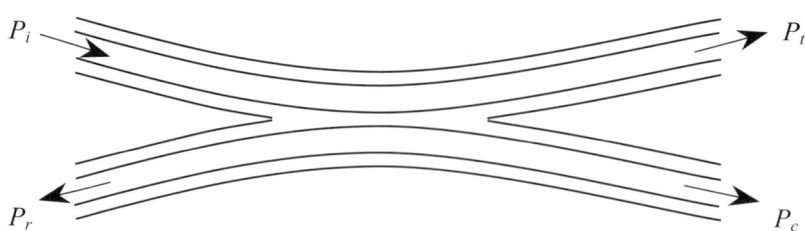

Figure 7.7. Fiber-optic directional coupler.

In practice, there are two important methods of fabricating fiber-optic couplers: polishing and fusing. The polishing method is based on exposing the core of the fiber by mechanically polishing off the cladding. The fiber first is bonded on a curved groove fabricated on a glass block. The groove depth at the center of the block is slightly greater than the cladding diameter of the fiber, as shown in Figure 7.8(a). The block along with the fiber is then ground and polished by standard mechanical polishing techniques until the core is almost exposed. Two such polished blocks are finally combined to form a polished fiber coupler, as shown in Figure 7.8(b).

A useful method to verify the closeness of the core to the polished surface is to observe the change in the light transmission through the fiber when a drop of index-matching liquid with a refractive index slightly greater than the core is spread on the polished region. If the core is far from the polished surface, there will be almost no change in light transmission. If the core is very close to the polished surface, however, the light transmission through the fiber will drop rapidly because of leakage of light from the core to the higher-refractive-index liquid. According to the drop in power, the closeness of the core to the polished surface can be estimated.

If the space between the two blocks is filled with index-matching liquid, we can laterally move one block with respect to the other. This will change the fiber core separation and in turn change the coupling ratio. Such polished fiber couplers are called tunable polished fiber couplers. The coupling ratio of commercial tunable polished fiber couplers can be tuned continuously from 0 to 100. The insertion loss of such couplers can be as low as 0.005 dB. The excess loss of the couplers can be as low as 0.05 dB. The directivity of the couplers can be as small as −70 dB.

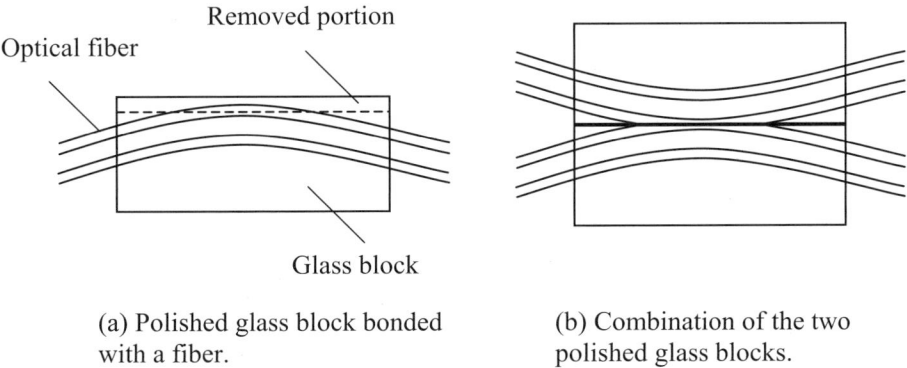

(a) Polished glass block bonded with a fiber.

(b) Combination of the two polished glass blocks.

Figure 7.8. Fabrication of polished fiber-optic directional couplers.

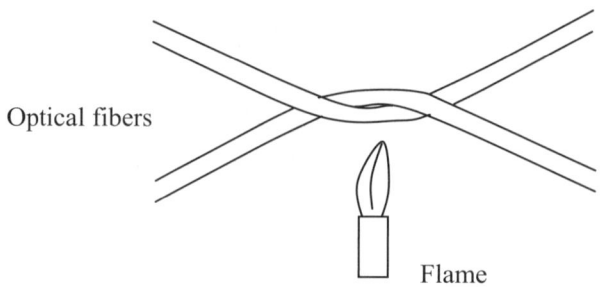

Optical fibers

Flame

Figure 7.9 Fabrication of fused fiber-optic directional couplers.

Moreover, by using polarization-maintaining fibers, we can fabricate polarization-maintaining fiber couplers or polarization-splitting fiber couplers. To
fabricate or utilize such fiber couplers, it is very important to achieve correct
alignment of the principal axes of the polarization-maintaining fibers.

Polished fiber couplers have excellent characteristics, but fabricating them
is a time-consuming process. In contrast, fused fiber couplers are easier to fabricate. Fused couplers are fabricated by first removing their plastic jackets,
slightly twisting two bared single-mode fibers, and then heating and pulling
them so that the fibers fuse laterally with each other, as shown in Figure 7.9.
The light power exiting from a fiber generally oscillates as a function of the
drawn length, and the variation of the output power from another fiber is com-
plementary. Therefore, carefully monitoring the exiting light power from a fiber during the heating and pulling process, we can make a fused fiber coupler
with any desired coupling ratio.

According to the number of leading fibers, fiber-optic couplers can also be
categorized into 1×2 Y-type couplers, 2×2 X-type couplers, and M×N multiple-
port couplers, as shown in Figure 7.10. Y-type couplers can be made out of X-
type couplers by cutting one of their fiber arms. Under some circumstances,
however, the reflection from the dead end of such Y-type fiber-optic couplers
may need to be considered.

It should be noted that the coupling ratio of a fiber-optic coupler might
vary as the wavelength changes. In wavelength-independent couplers, the differences are negligibly small. In wavelength-dependent couplers, however, the
differences can be made very large. Wavelength-selective couplers are designed to separate the light beams of different wavelengths and couple them
into the different fiber ports. The major use of such wavelength-selective couplers is in wavelength-division multiplexing fiber-optic systems, where two or
more signals are sent into a single fiber at different wavelengths and later they
are separated from each other by using a wavelength-selective coupler.

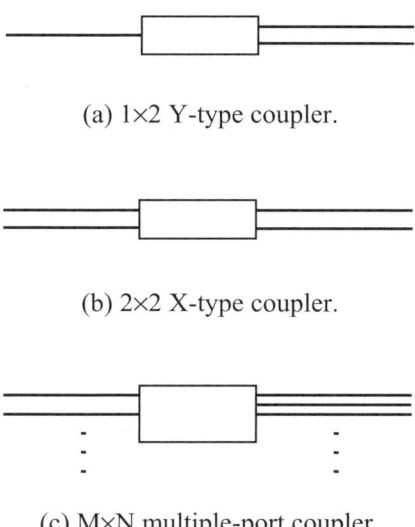

(a) 1×2 Y-type coupler.

(b) 2×2 X-type coupler.

(c) M×N multiple-port coupler.

Figure 7.10. Classification of fiber-optic couplers.

7.2.2 Fiber-optic Polarizers

In order to reduce the polarization noise and improve the quality of the effective signal, it is frequently necessary to polarize the light propagating in an optical fiber. So far, a number of fiber-optic polarizers have been proposed, but most of them are still being developed. In this subsection, we introduce two important ones: the crystal-clad polarizer and the metal-clad polarizer.

Crystal-clad polarizers use the evanescent field in the fiber cladding to couple the unwanted polarization wave away from the fiber [6]. Part of the cladding is removed over a small length of single-mode fiber to allow access to the evanescent field, and a birefringent crystal is used to replace the removed portion of cladding, as shown in Figure 7.11. If the refractive index of the crystal is greater than the effective refractive index of the optical fiber, the guided wave will excite a transmission wave in the crystal and therefore light will escape from the fiber. On the other hand, if the refractive index of the crystal is less than the effective refractive index of the fiber, no transmission wave will be excited and no light will escape from the fiber. Hence, by properly choosing the crystal and fiber, we can make one polarization mode radiate while the other polarization mode remains guided. With this principle, fiber-optic polarizers with very low extinction ratios (–60 dB) have been demonstrated. (The extinction ratio is defined as the ratio of the radiating mode transmittance and the guided mode transmittance in decibels).

Birefringent crystal

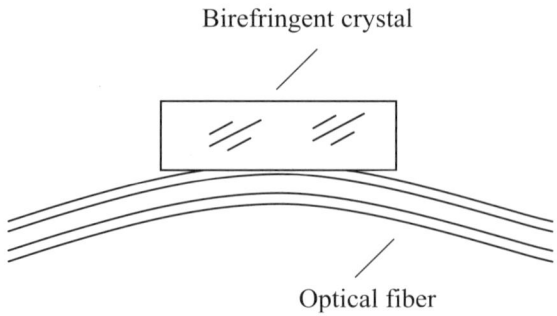

Optical fiber

Figure 7.11 Crystal-clad fiber-optic polarizer.

Metal-clad polarizers are based on the fact that a dielectric-metal interface can support a guided TM wave (or surface plasmon mode), which is polarized perpendicular to the dielectric-metal interface. Because a metal has an imaginary component of dielectric constant, such a surface plasmon mode is lossy. On the other hand, however, the dielectric-metal interface cannot support any guided TE wave because it cannot satisfy the required boundary conditions at the dielectric-metal interface. If part of the cladding is removed over a small length of single-mode fiber and the removed side of the fiber is coated with a metal, the light with a polarization direction perpendicular to the dielectric-metal interface (TM wave) will suffer much loss due to the metal, whereas the light with a polarization direction parallel to the dielectric-metal interface (TE wave) will suffer much less loss. For a metal-clad polarizer of D-type fiber, for instance, an extinction ratio of −39 dB can be achieved [36].

7.2.3 Fiber-Optic Polarization Controllers

We know that single-mode optical fibers are not birefringent (i.e., the two orthogonal polarization modes in single-mode optical fibers have the same effective refractive index). Bending such a fiber will introduce stress in the fiber and will make the fiber linearly birefringent. The bending-induced birefringence in a single-mode fiber can be determined by

$$\Delta n_e = n_{ex} - n_{ey}$$

$$= -C \left(\frac{b}{R} \right)^2 , \qquad (7\text{-}25)$$

where n_{ex} and n_{ey} represent the effective refractive indices of the two modes polarized in the plane and perpendicular to the plane of the bend fiber, respectively, b is the outer radius of the fiber, R is the radius of the bend fiber, and C is a constant that depends on the elasto-optic property of the fiber (for silica fibers, $C \approx 0.133$ at 633 nm). Notice that a smaller loop radius can produce a larger birefringence, but it will also introduce a bigger attenuation. Therefore, the very small bend radius is not practical.

The bend-induced linear birefringence can be used to make an in-line fiber-optic polarization controller [7], as shown in Figure 7.12(a). The polarization controller consists of three fiber loops: the first and the last fiber loops act as quarter-wave retarders and the central fiber loop acts as a half-wave wave plate. The fiber is fixed at points marked A, B, C, and D, and the three fiber loops are free to rotate, as shown in Figure 7.12(b). Rotation of each fiber loop will rotate the principal axes of the birefringent fiber section with respect to the input polarization state. This is analogous to rotation of a conventional bulk half-wave or a quarter-wave plate with respect to the incident light. Rotation of the three fiber loops is equivalent to rotation of a combination of the $\lambda/4$, $\lambda/2$, and $\lambda/4$ wave plates, and therefore any input polarization state can be transformed to any other output polarization state.

Fiber-optic polarization controllers are frequently used in fiber-optic interferometers containing long lengths of single-mode fibers, such as single-mode fiber-optic gyroscopes. Under these circumstances, the polarization states of the interfering beams must be exactly the same, so that the polarization noise can be reduced and the best signal-to-noise ratio can be achieved.

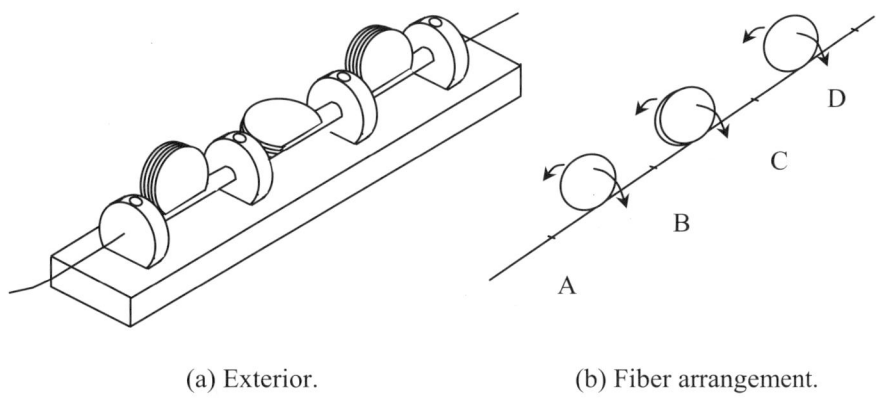

(a) Exterior.   (b) Fiber arrangement.

Figure 7.12. An in-line fiber-optic polarization controller. (After [7].)

7.2.4 Fiber-optic Connectors

Fiber-optic connectors (or fiber connectors) are used for temporarily connecting two optical fibers together. A number of demountable optical fiber connectors have been developed, and the average insertion losses are in the range from 0.2 to 3 dB. Fiber connectors may be categorized into two groups, butt-jointed connectors and expended-beam connectors. Figure 7.13 shows a butt-jointed connector that employs a V-shaped groove fabricated on a stainless steel substrate to align the two bared fiber ends and employs two springs, one on each end of the substrate, to hold the connected fibers. The aligned fiber ends usually are fixed by means of a cover plate.

Expanded-beam connectors employ coupling lenses or gradient-index lenses on the ends of the two fibers, as shown in Figure 7.14. One lens is used to collimate the light emerging from the transmitting fiber, while another lens is used to focus the expanded beam onto the core of the receiving fiber. The advantage of this type of connector is that, since the beam is collimated, separation of the fiber ends may take place within the connector. Thus, the connector is less dependent on lateral alignment. In addition, other optical processing elements, such as beam splitters, polarizers, or optical isolators, can easily be inserted into the expanded beam.

Gradient-index lenses are cylindrical glass rods whose optical refractive index decreases radially outward from the central axis. Since the optical refractive index in the central region is larger than that in the outside region, the meridional rays will periodically bend to the central axis along with the light propagation. Theoretical analysis indicates that, if a gradient-index lens has a parabolic index profile,

$$n(r)^2 = n_0^{\ 2}\left(\frac{1 - a^2 r^2}{2}\right),$$

$$(7\text{-}26)$$

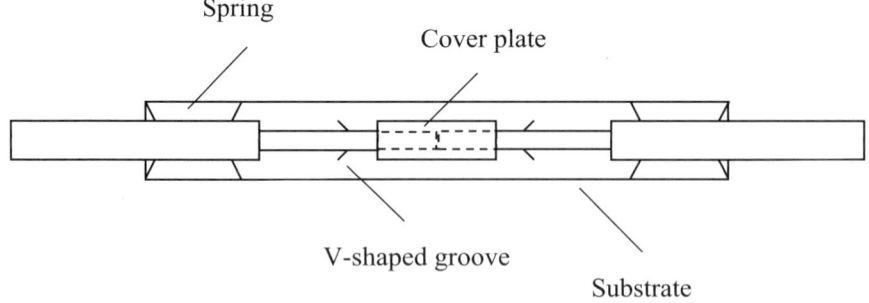

Figure 7.13. Butt-jointed connector.

Coupling lenses

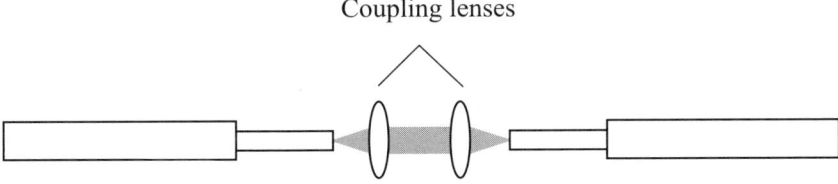

(a) Expanded-beam connector with collimating lens.

Gradient-index lenses

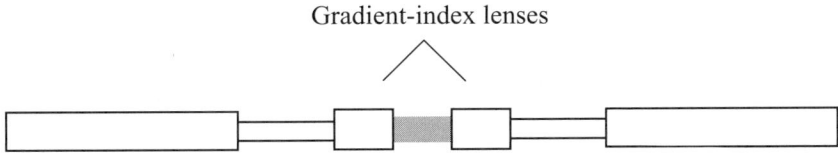

(b) Expanded-beam connector with gradient-index lens.

Figure 7.14. Expanded-beam connectors.

where n_0 is the refractive index on the central axis, a is a constant, and r is the radial coordinate, the meridional rays starting from a point on the central axis will focus on another point on the axis, as shown in Figure 7.15(a). The period of the rays on the central axis is called a pitch, and one pitch P equals

$$P = \frac{2\pi}{a}.$$

(7-27)

The most frequently used gradient-index lenses are of 1/4 pitch (or slightly less than 1/4 pitch) or 1/2 pitch (or slightly less than 1/2 pitch). Quarter-pitch gradient-index lenses, like collimating lenses, can collimate the light emerging from a point or focus an expanded beam onto a point, as shown in Figure 7.15(b). Half-pitch gradient-index lenses, like projection lenses, are generally used to form an image of the light source near the front face onto the back face, as shown in Figure 7.15(c).

Gradient-index lenses have been widely used in fiber-optic systems because they are small, their cylindrical shape makes them easy to mount and align, and their planar end faces are easy to attach to fibers. Figure 7.16 shows two other types of fiber-optic connectors that are frequently used for laser-to-fiber connections. Note that, because the second fiber-optic connector contains an optical isolator, the emerging light actually is polarized.

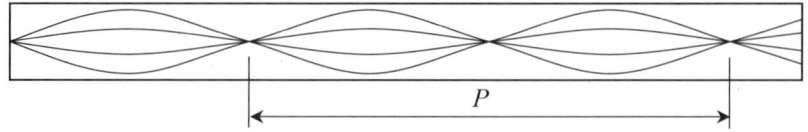

(a) Propagation of light in a gradient-index glass rod.

(b) Quarter-pitch gradient-index lenses.

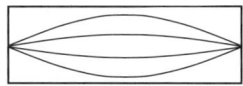

(c) Half-pitch gradient-index lenses.

Figure 7.15. Optical gradient-index lenses

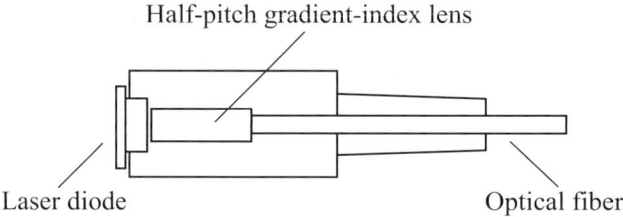

(a) Without an optical isolator.

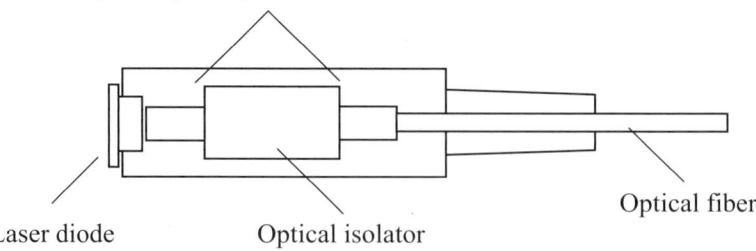

(b) With an optical isolator.

Figure 7.16. Fiber-optic connectors for laser-to-fiber connections.

7.2.5 Fiber-optic Splices

Fiber-optic splices (or fiber splices) are used for connecting two fibers together permanently. Two commonly used fiber splices are fusion splices and V-groove splices.

Fusion splices are made by thermally bonding the prepared fiber ends together. The fiber ends are first butted and held in alignment by mechanical micropositioners. The butted joint is then heated with an electric arc so that the fiber ends are momentarily melted and therefore bonded together, as shown in Figure 7.17(a). This method can produce a very low splice loss, typically averaging less than 0.09 dB. However, care must be taken with this method since misalignment of fiber ends, surface damage due to handling, surface defects created during heating, and residual stress due to changes in chemical composition after the material melts can produce a poor splice. In addition, the heat used to fuse the fibers may make the fiber in the close vicinity of the splice extremely brittle. Therefore, the fused splice is usually sealed in a protective tube, as shown in Figure 7.17(b).

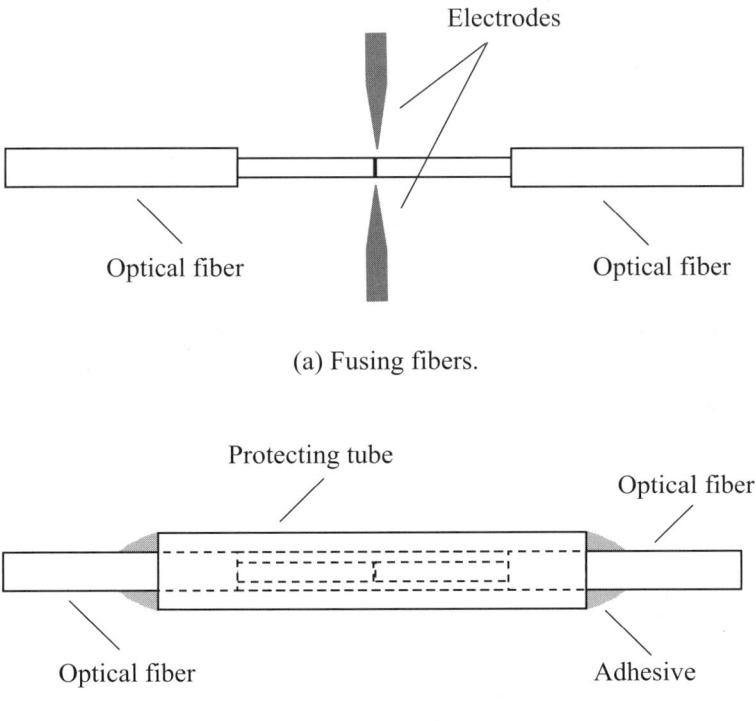

(a) Fusing fibers.

(b) Splice sealing.

Figure 7.17. Fusion splicing of optical fibers.

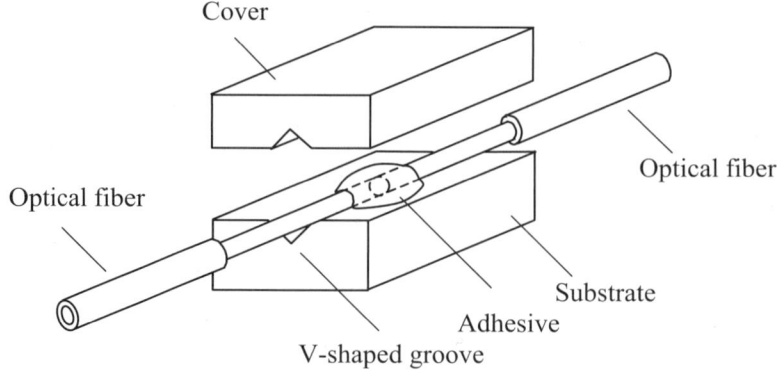

Figure 7.18. V-groove splicing of optical fibers.

To make V-groove splices, the prepared fiber ends are first butted together in a V-shaped groove, as shown in Figure 7.18. They are then bonded together with an adhesive or sealed with a cover. The V-shaped groove and cover can be made on a block of either glass or plastic. The splice loss in this method depends strongly on the fiber size and eccentricity (the position of the core relative to the center of the fiber).

Optical fibers and fiber-optic components are a broad topic. The detailed discussion about them is beyond the scope of this book. Fortunately, we have what we need for constructing fiber-optic FMCW interferometers. It should be announced that because some fiber-optic components (such as fiber-optic polarizers, fiber-optic isolators, and polarization-maintaining fiber-optic couplers) have not yet been widely used due to their cost, in this book we still use the corresponding bulk optical elements. The benefits of this are that the fiber-optic systems are easy to explain in principle, are close to the configurations originally reported, and are easy to reconstruct with common laboratory equipment.

7.3 Fiber-optic Michelson FMCW Interferometers

Figure 7.19 schematically shows a fiber-optic Michelson FMCW interferometer, which simply consists of an X-type single-mode fiber-optic coupler (FC) and two mirrors (M_1 and M_2) [18]. An FMCW laser beam is launched into the first input fibers of FC, divided into two beams propagating along the two output fibers. These two beams are reflected by M_1 and M_2 (usually the cleaved ends of the fibers are coated with silver and thus act as mirrors) and propagate back to FC to mix coherently. The beat signal produced propagates along the second input fiber of FC and finally is detected by a photodetector.

The optical path difference OPD between the two interfering beams can be written as

$$OPD = 2n_e(l_2 - l_1),$$

(7-28)

where n_e is the effective refractive index of the single-mode fiber and l_1 and l_2 are the lengths of the two output fibers of FC.

If the frequency of the laser is modulated with a sawtooth waveform, for instance, the intensity of the beat signal detected in a modulation period $I(t)$ can be written as

$$I(OPD,t) = I_0\left[1 + V\cos\left(\frac{2\pi\Delta v v_m OPD}{c}t + \frac{2\pi}{\lambda_0}OPD\right)\right]$$

$$= I_0[1 + V\cos(2\pi v_b t + \phi_{b0})]$$

(2-63)

where I_0 is the average intensity of the beat signal, V is the contrast of the beat signal, Δv is the optical frequency modulation excursion, v_m is the modulation frequency, c is the speed of light in free space, λ_0 is the central optical wavelength, and v_b and ϕ_{b0} are the frequency and initial phase of the beat signal, respectively. Considering Equation (7-28), the frequency of the beat signal v_b can be written as

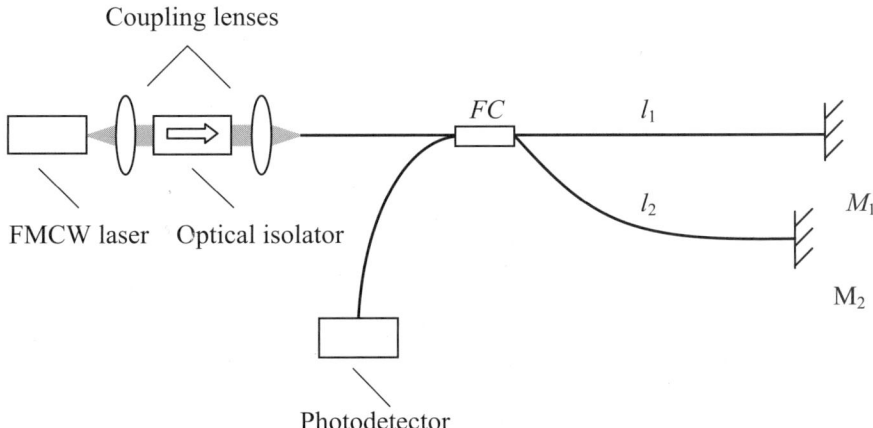

Figure 7.19 The fiber-optic Michelson FMCW interferometer. (After [18].)

$$v_b = \frac{2n(l_2 - l_1)\Delta v v_m}{c} ,$$

(7-29)

and the initial phase of the beat signal ϕ_{b0} can be written as

$$\phi_{b0} = \frac{4\pi n(l_2 - l_1)}{\lambda_0} ,$$

(7-30)

Hence, measuring the frequency of the beat signal, we can obtain the absolute value of the length difference between the two fiber arms ($l_2 - l_1$)

$$l_2 - l_1 = \frac{c}{2n\Delta v v_m} v_b ,$$

(7-31)

and measuring the phase shift of the beat signal, we can obtain the elongation of a fiber arm Δl_2,

$$\Delta l_2 = \frac{\lambda_0}{4\pi n_e} \Delta\phi_{b0} ,$$

(7-32)

where l_1 is assumed to be a constant.

If one of the output fibers of the fiber coupler is placed in a strain or temperature field, the optical path difference will vary in sympathy with the strain or temperature. Therefore, this fiber-optic interferometer can be used for measuring strain or temperature.

Figure 7.20 shows another version of the fiber-optic Michelson FMCW interferometer, in which the signal arm of the interferometer consists of a length of output fiber l_1, a collimating lens (L), and a mirror (M_1). The optical path difference OPD between the two interfering beams in this case can be written as

$$OPD = n_e l_2 - [n_e l_1 + n l_1' + (n_L - n)d] ,$$

(7-33)

where n is the refractive index of air, n_e is the effective refractive index of the single-mode fiber, n_L is the refractive index of L, l_1 and l_2 are the lengths of the output fibers of the fiber coupler, l_1' is the distance from the end of the output fiber to M_1, and d is the thickness of L.

Obviously, this type of fiber-optic Michelson FMCW interferometer is suited to measuring the distance, displacement, or speed of an object.

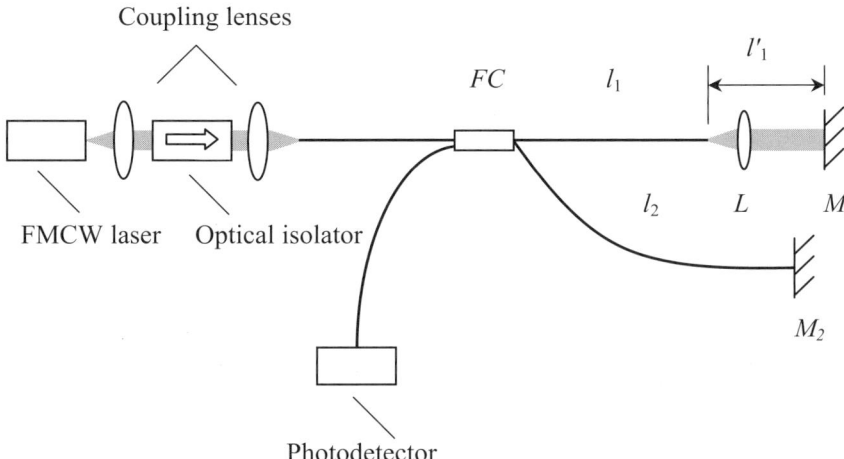

Figure 7.20. The fiber-optic Michelson FMCW interferometer with a fiber-air-mixed arm.

7.4 Fiber-optic Mach-Zehnder FMCW Interferometers

The fiber-optic Mach-Zehnder FMCW interferometer in Figure 7.21 consists of two Y-type single-mode fiber couplers (FC_1 and FC_2) [20]. The two output fibers of FC_1 with different lengths are connected with the two input fibers of FC_2 to construct an unbalanced fiber-arm interferometer. An FMCW laser beam is passed through FC_1, which splits the beam equally into the two output fiber arms. After traversing the fiber lengths, these two beams are recombined

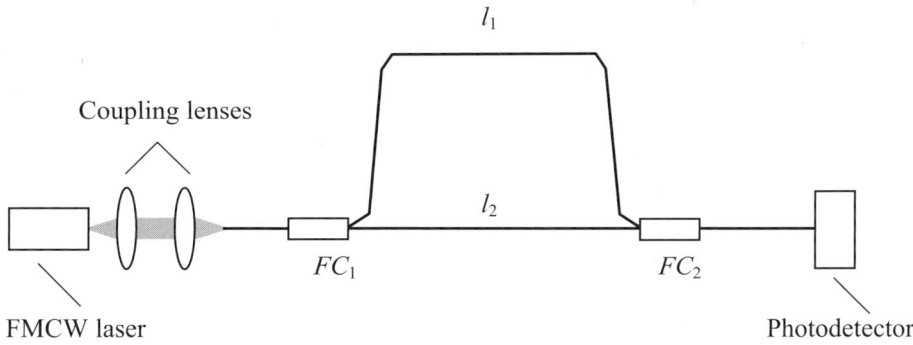

Figure 7.21. The fiber-optic Mach-Zehnder FMCW interferometer. (After [20].)

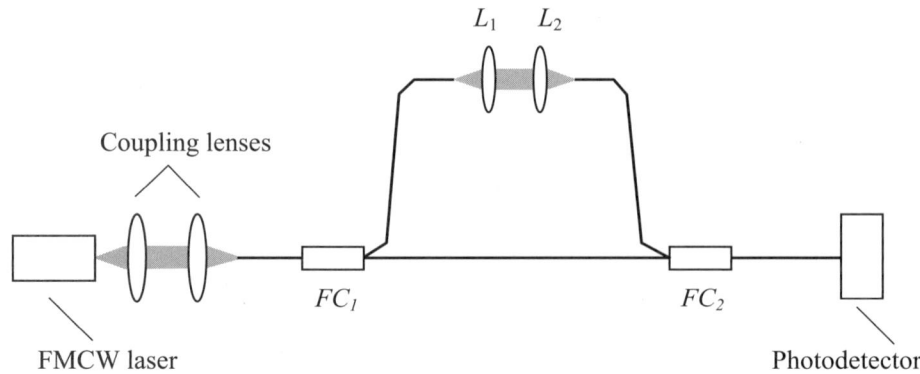

Figure 7.22. The fiber-optic Mach-Zehnder FMCW interferometer with a broken fiber arm. (After [47].)

by FC_2. The beat signal produced propagates along the output fiber of FC_2 and is finally detected by a photodetector. The optical path difference OPD between the two interfering beams can be written as

$$OPD = n_e(l_2 - l_1),$$

$$(7\text{-}34)$$

where n_e is the effective refractive index of the single-mode fiber, and l_1 and l_2 are the lengths of the two fiber arms. The signal analysis for the fiber-optic Mach-Zehnder FMCW interferometer is similar to that for the fiber-optic Michelson FMCW interferometer; therefore, we do not discuss it here.

Figure 7.22 shows another version of the fiber-optic Mach-Zehnder FMCW interferometer, in which one of the fiber arms is broken and two collimating lenses (L_1 and L_2) are inserted in the fiber arm [47]. Obviously, the length of this arm can be changed over a long range.

7.5 Fiber-optic Fabry-Perot FMCW Interferometers

The fiber-optic Fabry-Perot FMCW interferometer, as shown in Figure 7.23, consists of a Y-type single-mode fiber-optic coupler (FC) linked to a fiber cavity (a short length of single-mode fiber) by using a fiber connector (FN) [51]. An FMCW laser beam passes through the fiber coupler and is partially reflected by the end of the output fiber of FC. (Usually the cleaved end of the fiber is coated with a partially reflecting film.) The remainder of the beam, however, propagates along the fiber cavity and then is reflected by a mirror

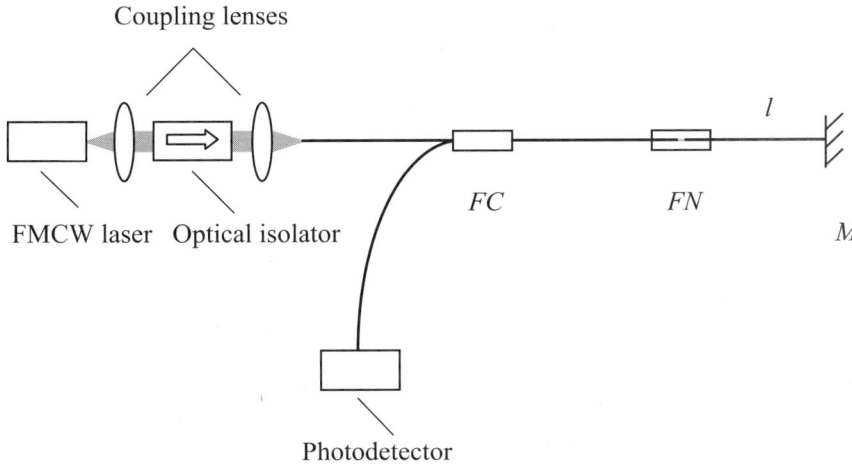

Figure 7.23. The fiber-optic Fabry-Perot FMCW interferometer. (After [51].)

(*M*). The two reflected beams mix coherently in the fiber coupler, and the beat signal produced is detected by a photodetector.

The optical path difference *OPD* between the two reflected beams can be written as

$$OPD = 2n_e l,$$ (7-35)

where n_e is the effective refractive index of the single-mode fiber, and *l* is the length of the single-mode fiber cavity. The signal analysis for the fiber-optic Mach-Zehnder FMCW interferometer is similar to that for the fiber-optic Michelson FMCW interferometer.

Figure 7.24 shows another version of the fiber-optic Fabry-Perot FMCW interferometer, in which a collimating lens (*L*) and a length of air path are used rather than the fiber cavity. The optical path difference *OPD* between the two reflected beams in this case can be written as

$$OPD = 2[nl + (n_L - n)d],$$ (7-36)

where *l* is the distance from the fiber end to *M*, *n* is the refractive index of air, and n_L and *d* are the refractive index and thickness of the collimating lens, respectively.

The advantage of fiber-optic Fabry-Perot FMCW interferometers is that they are free from environmental interference because the two interfering beams always propagate along the same fiber in the same mode except for the

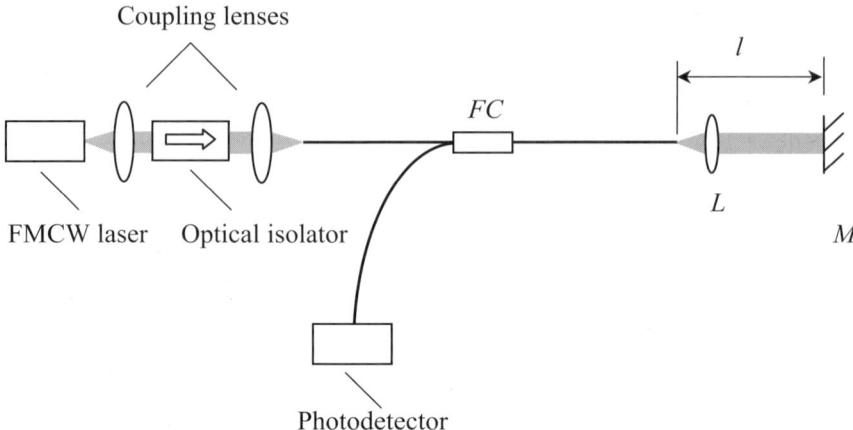

Figure 7.24. The fiber-optic Fabry-Perot FMCW interferometer with a fiber-air-mixed arm.

region of the fiber cavity or air cavity, with the result that the effects on the two interfering beams are naturally compensated. Therefore, the fiber-optic Fabry-Perot FMCW interferometers are suitable for remote measurement of the parameters of a distant target.

Chapter 8

Multiplexed Fiber-optic Frequency-Modulated Continuous-Wave Interferometers

An additional important advantage of fiber-optic interferometers is that they can be combined to form fiber-optic interferometer networks—the *multiplexed fiber-optic interferometers*. A multiplexed fiber-optic interferometer usually uses a single optical source and a single photodetector, but it can simultaneously measure a number of different targets or different parameters so that the cost per individual interferometer can be reduced. Another important application of the multiplexed fiber-optic interferometer is that one or more individual interferometers in the network can be used to measure the effect of surrounding conditions on the network itself, so that the error introduced by the environment can be dynamically compensated and the performance of the multiplexed fiber-optic interferometer can be significantly improved.

To make a multiplexed fiber-optic interferometer, a suitable signal modulation-demodulation method and a suitable topological arrangement of fiber-optic interferometers generally are required in order that the information from each individual fiber-optic interferometer can be separated.

In this chapter, we will discuss some important multiplexing methods, including the frequency-division multiplexing method, time-division multiplexing method, time-frequency-division multiplexing method, and coherence-division multiplexing method, and the corresponding multiplexed fiber-optic FMCW interferometers.

The polarization-division and intensity-division multiplexing methods are also frequently used in practice. However, since these two methods are relatively simple in principle, we will discuss them when we use them in the next chapter.

8.1 Frequency-Division Multiplexed Fiber-optic FMCW Interferometers

The frequency-division multiplexing method is based on the fact that the frequency of the beat signal from an FMCW interferometer depends on the optical path difference between the two interfering beams in the interferometer. If the optical path differences in a number of interferometers are different from each other, the beat signals from these interferometers will be different in frequency, and therefore they can be separated by using electric band-pass filters.

Figure 8.1 shows schematically a frequency-division multiplexed fiber-optic FMCW interferometer, which consists of a number of fiber-optic Mach-Zehnder FMCW interferometers [34]. The interferometers are connected in parallel between two leading fibers (usually called the input fiber bus and output fiber bus, respectively) with fiber-optic couplers, and they are constructed with different optical path differences.

Another requirement for achieving a frequency-division multiplexed fiber-optic FMCW interferometer is that the laser beams from different interferometers should be incoherent. Otherwise, a multiple-beam interferometric signal will be produced, which, of course, is hard to separate into the signals of individual interferometers, as indicated in Section 2.3. A simple way to avoid this trouble is to make the connecting fiber (represented by a circle in Figure 8.1) between interferometers longer than the coherence length of the laser (neglecting the fiber length of the interferometers). Under this condition, the light beams

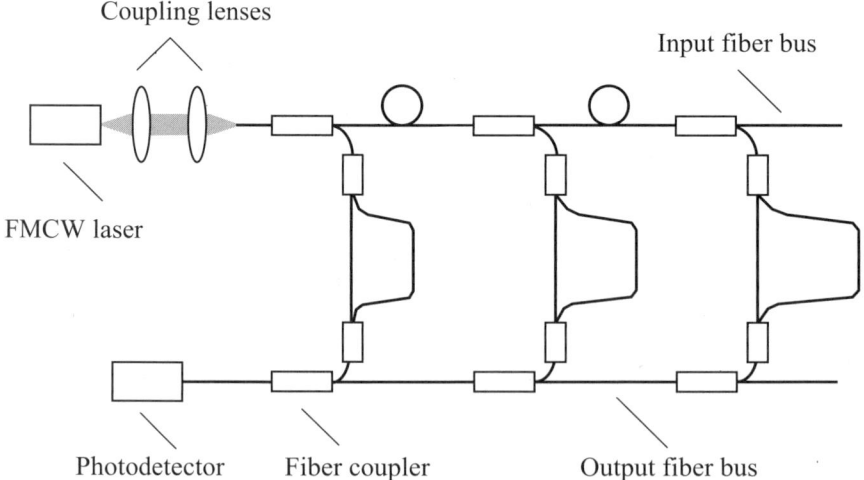

Figure 8.1. Frequency-division multiplexed fiber-optic Mach-Zehnder FMCW interferometer. (After [34].)

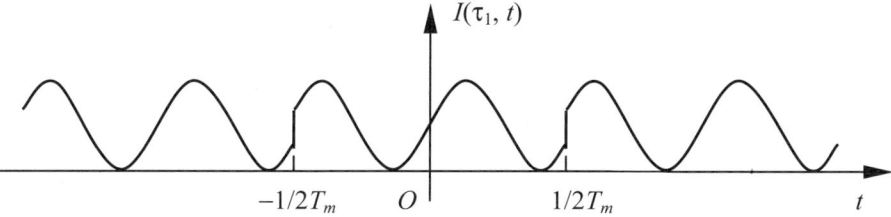

(a) Signal from the first interferometer.

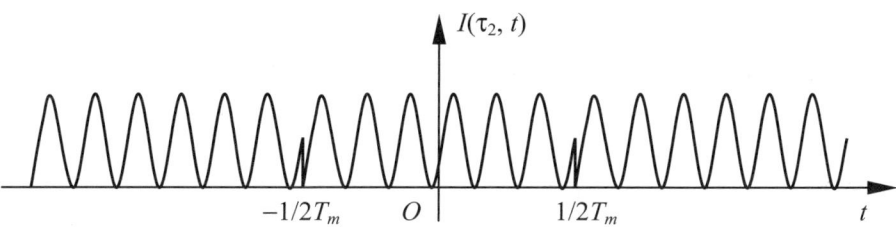

(b) Signal from the second interferometer.

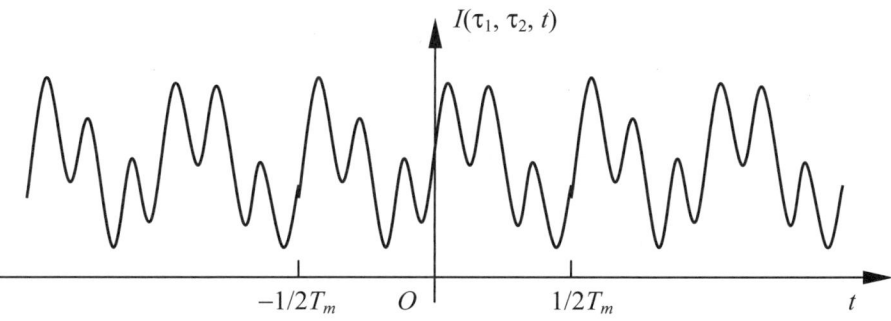

(c) Signal from the multiplexed interferometer.

Figure 8.2. Waveforms of the signals from a double-interferometer frequency-division multiplexed fiber-optic FMCW interferometer.

from different interferometers are no longer coherent, and the detected signal from the whole fiber-optic network will be equal to the sum of the intensities of the beat signals from each individual interferometer, as shown in Figure 8.2.

The frequency-division multiplexed fiber-optic FMCW interferometer can also be constructed with fiber-optic Michelson FMCW interferometers, as shown in Figure 8.3. An FMCW laser beam is launched into the first fiber-optic coupler. The output fiber of the coupler is used as a fiber bus and connected to a number of fiber-optic Michelson FMCW interferometers. In the

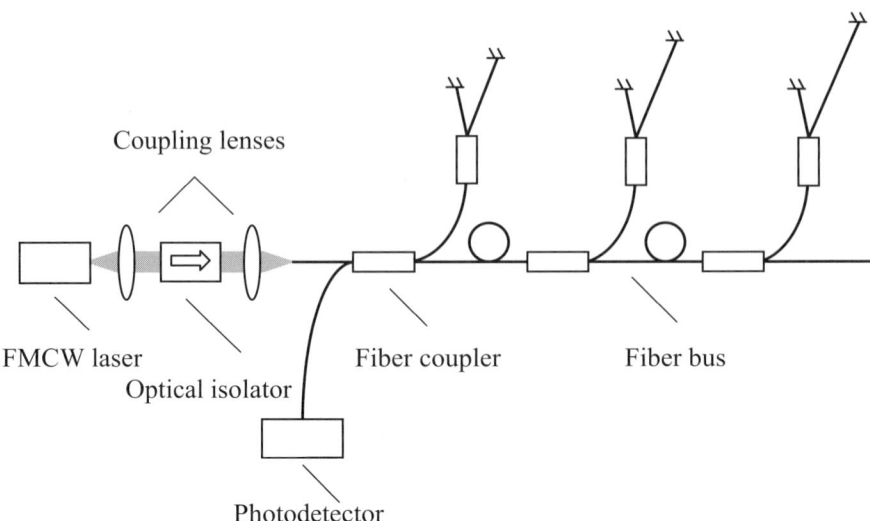

Figure 8.3. Frequency-division multiplexed fiber-optic Michelson FMCW interferometer.

same way, the optical path differences in these interferometers are required to be different from each other, and the connecting fibers are required to be longer than half of the coherence length of the laser. The beat signals produced by each individual interferometer propagate back to the first coupler, emerge from the second input fiber of the coupler, and are detected by a photodetector.

The drawbacks of the frequency-division multiplexing method include:

(1) Because each interferometer has to work at a specified frequency, the variation range of the delay time or *OPD* of each individual interferometer is restricted.

(2) Because the beat signal from each individual interferometer generally contains a number of harmonics, there is signal cross talk between the interferometers (see Section 9.2).

8.2 Time-Division Multiplexed Fiber-optic FMCW Interferometers

The time-division multiplexing method is based on the gated frequency modulation, for instance, the gated sawtooth-wave modulation [42]. A number of fiber-optic Mach-Zehnder FMCW interferometers are connected in parallel between the input fiber bus and the output fiber bus with fiber-optic couplers and long lengths of connecting fibers (usually called delay fibers, represented by a double circle), as shown in Figure 8.4. The delay time introduced by each delay fiber is required to be longer than the frequency modulation period T_m, and the total delay time introduced by the whole fiber-optic network is made to be shorter than the separation of the gated sawtooth T_m'. Under these conditions, the beat signals from each interferometer arrive at the photodetector at different times without overlapping, and therefore, they can be separated using an electric gate circuit.

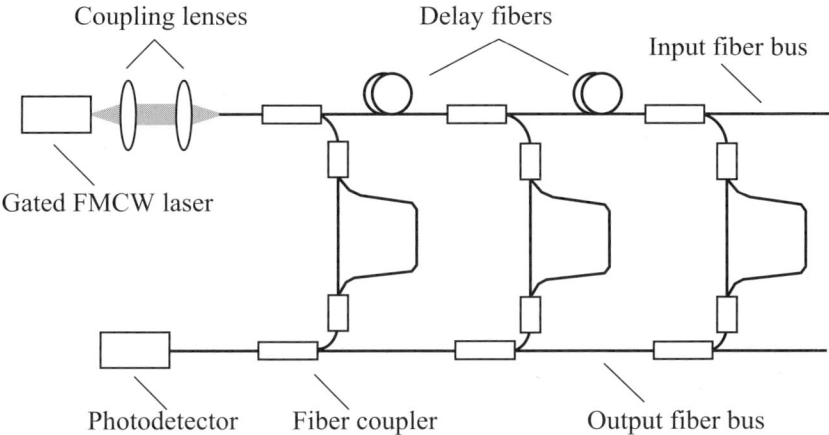

(a) Configuration of the interferometer.

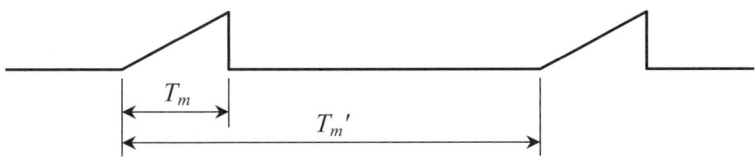

(b) Waveform of the modulation signal.

(c) Waveform of the beat signal.

Figure 8.4. Time-division multiplexed fiber-optic Mach-Zehnder FMCW interferometer. (After [42].)

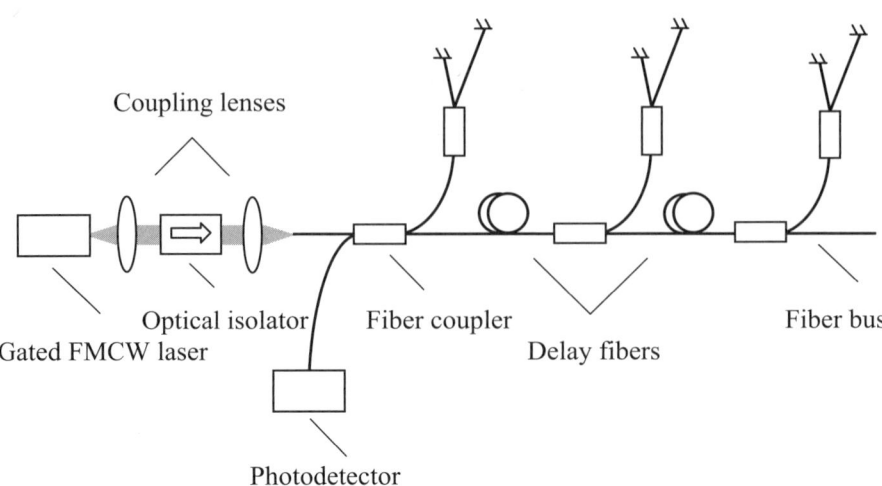

Figure 8.5. Time-division multiplexed fiber-optic Michelson FMCW interferometer.

Figure 8.5 shows schematically another time-division multiplexed fiber-optic FMCW interferometer, which consists of a number of fiber-optic Michelson FMCW interferometers. A gated FMCW laser beam is launched into the first fiber-optic coupler. The output fiber of this coupler, which is used as a fiber bus, is connected to a number of fiber-optic Michelson FMCW interferometers with a length of delay fiber. The delay time introduced by each delay fiber is required to be longer than half the sawtooth-wave modulation period T_m, and the total delay time introduced by the whole fiber-optic network is required to be shorter than half the separation the gated sawtooth T_m'.

The time-division multiplexing method can overcome the problems of the frequency-division multiplexing method. However, it requires long lengths of delay fibers, which not only increases the cost of the system but also introduces big attenuation, which limits the maximum number of multiplexed interferometers. For instance, if the modulation period is 5 μs, the length of each delay fiber in Figure 8.4 should be longer than 1 km, where the effective refractive index of the single-mode fiber is assumed to be 1.5.

8.3 Time-Frequency-Division Multiplexed Fiber-optic FMCW Interferometers

In order to increase the number of multiplexed interferometers and reduce the cross talk between the beat signals, a feasible way is to combine the time-division

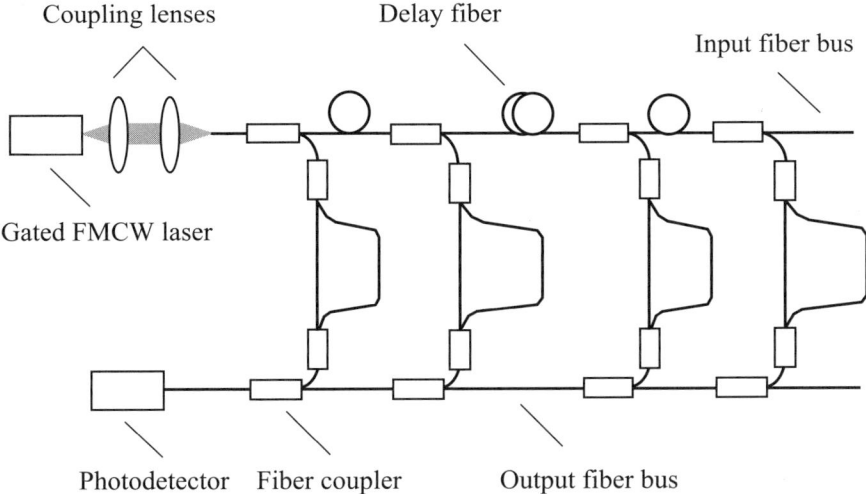

Figure 8.6. Time-frequency-division multiplexed fiber-optic Mach-Zehnder FMCW interferometer. (After [43].)

multiplexing method and the frequency-division multiplexing method [43].

Figure 8.6 illustrates a time-frequency-division multiplexed fiber-optic Mach-Zehnder FMCW interferometer, which consists of several subsystems. Each subsystem comprises a number of fiber-optic Mach-Zehnder interferometers with different optical path differences, so that they can be frequency-division multiplexed. Each subsystem is separated from its neighbor by a delay fiber, so that the subsystems can be time-division multiplexed.

Figure 8.7 shows another version of the time-frequency-division multiplexed FMCW interferometer, which consists of a number of fiber-optic Michelson interferometers and whose operating principle is the same as the previous one.

8.4 Coherence-Division Multiplexed Fiber-optic FMCW Interferometers

The fundamental idea behind coherence-division multiplexing is to use a number of interferometers with different optical path differences that are longer than the coherence length of the laser (usually called sensing interferometers) for measurement and use the same number of additional interferometers (usually called receiving interferometers) to balance the optical path differences in the sensing interferometers to collect information [25].

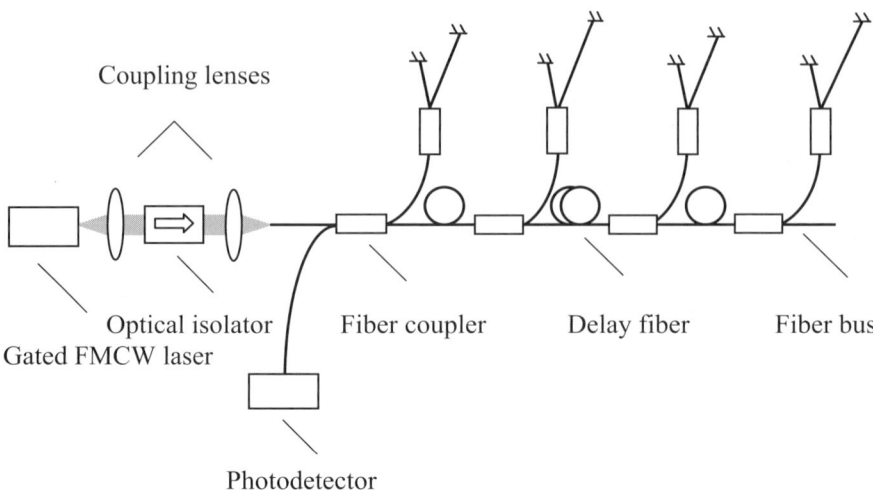

Figure 8.7. Time-frequency-division multiplexed fiber-optic Michelson FMCW interferometer.

Figure 8.8 shows schematically a coherence-division multiplexed fiber-optic Mach-Zehnder FMCW interferometer, in which the sensing interferometers are connected in parallel between the input fiber bus and output fiber bus and the receiving interferometers are attached to the output fiber bus separately. The optical path differences in the sensing interferometers are made to be different from each other and longer than the coherence length of the laser, so that the laser beams in each interferometer cannot properly interfere without a matched receiving interferometer. Similarly, the connecting fibers between the sensing interferometers should also be made longer than the coherence length of the laser in order to keep the beams from different interferometers from coherently interfering.

Under these conditions, no beat signal can be directly detected from the output fiber bus. However, if the optical path difference in a receiving interferometer is similar to that in a sensing interferometer, there will always be two beams (one propagating along the long arm of the sensing interferometer and the short arm of the receiving interferometer and another propagating along the short arm of the sensing interferometer and the long arm of the receiving interferometer) that can coherently interfere and produce a beat signal at the following photodetector. Therefore, the variation of the optical path difference in the sensing interferometer can be measured with a matching receiving interferometer. Because each sensing interferometer has a corresponding matching receiving interferometer, the information from any other sensing interferometer can be discovered in the same way.

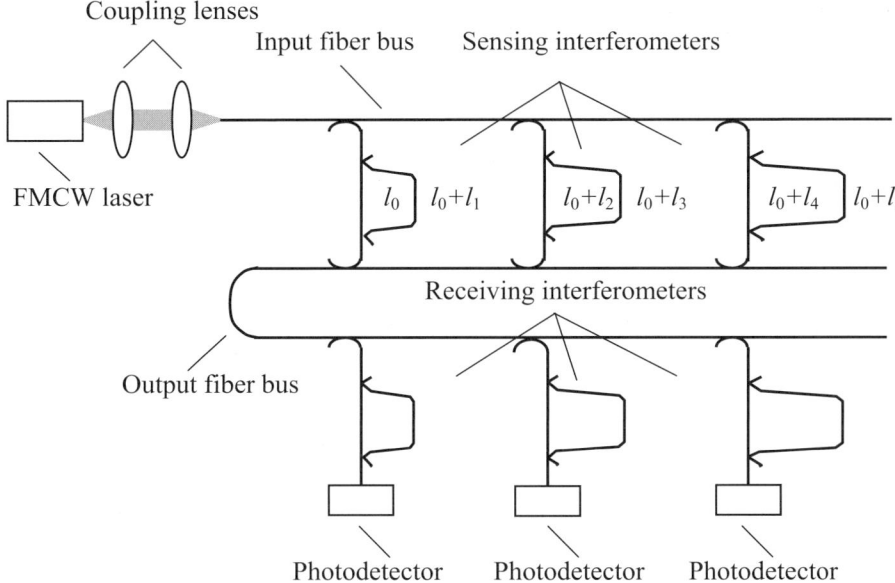

Figure 8.8 Coherence-division multiplexed fiber-optic Mach-Zehnder interferometer. (After [25].)

There is a potential risk that one receiving interferometer may match several pairs of fiber arms that belong to different sensing interferometers. However, this cross talk and ambiguity can be avoided if a proper fiber-optic arrangement is used. For instance, assuming that the coherence length of the laser is L_c and the different fiber lengths in sensing interferometers are l_1, l_2, ... l_k,..., as shown in Figure 8.8, the criterion for a coherence-division multiplexed interferometer without cross talk and ambiguity is

$$l_k \geq m_k L_c \qquad (k=1, 2, 3,...),\qquad (8\text{-}1)$$

where m_k is an integer that should satisfy

$$(m_k - m_j) \notin \{m_{k-i} - m_l\} \qquad (j<k,\ i<k,\ l<k-i).\qquad (8\text{-}2)$$

For instance, in the best case (where all the fiber arms have the shortest lengths), m_k should use the sequence (1, 3, 7, 12, 20, 30, 44, ...).

The maximum number of sensing interferometers in the coherence-division multiplexed FMCW fiber-optic interferometer is determined by the signal-noise ratio, which is related to the laser-power budget and the intensity noises from all sensing interferometers.

Chapter 9

Fiber-optic Frequency-Modulated Continuous-Wave Interferometric Sensors

Fiber-optic sensors are the twin technology of fiber-optic communications. Application of optical fibers to optical sensing is based on the fact that various properties of the light propagating through an optical fiber can be varied in sympathy with environmental parameters. Since the 1980s, a broad exploration in this field has led to many types of fiber-optic sensors, such as fiber-optic displacement sensors, fiber-optic strain sensors, fiber-optic stress sensors, fiber-optic temperature sensors, fiber-optic rotation sensors (i.e., fiber-optic gyroscopes), fiber-optic electric current sensors, fiber-optic magnetic field sensors, and fiber-optic chemical sensors. The areas employing fiber-optic sensors cover aerospace, marine, nuclear engineering, mechanical engineering, chemical engineering, civil construction, and environmental protection.

In this chapter, we will first introduce some basic knowledge about fiber-optic sensors and then present some fiber-optic sensors based on the principles of optical FMCW interference, including fiber-optic FMCW interferometric displacement sensors, fiber-optic FMCW interferometric strain sensors, fiber-optic FMCW interferometric stress sensors, fiber-optic FMCW interferometric temperature sensors, and fiber-optic FMCW interferometric rotation sensors (gyroscopes).

9.1 Introduction to Fiber-optic Sensors

Fiber-optic sensors employ light propagating through an optical fiber to detect an environmental parameter. In principle, any property of the light, such as intensity, color, frequency, phase, or polarization state, can be used to sense a physical or chemical parameter if this parameter affects any property of the light.

Fiber-optic sensors can be categorized as extrinsic fiber-optic sensors and intrinsic fiber-optic sensors. In an *extrinsic fiber-optic sensor*, the optical fiber is not directly affected by the parameter, while in an *intrinsic fiber-optic sensor*, the optical fiber experiences it directly.

Fiber-optic sensors have been shown to possess a number of advantages over conventional electrical sensors, such as immunity from electromagnetic interference, electrical isolation, chemical passivity, higher sensitivity, and being lightweight and versatile. The most important feature of fiber-optic sensors is that they can be modified or combined to create fiber-optic sensing networks—multiplexed fiber-optic sensors or distributed fiber-optic sensors. This feature is very important in practice since it enables information to be obtained from a single sensor, which otherwise would require several sensors. Another important application of the multiplexed fiber-optic sensor is that one or more individual sensors in the network can be used to measure the effect of the surrounding environment on the network itself so that the error introduced by environmental conditions can be dynamically compensated and the accuracy and long-term stability of the sensor can be significantly improved.

The *multiplexed fiber-optic sensor* is a sensor network that consists of several individual fiber-optic sensors and generally uses a single light source and a single photodetector. To create a multiplexed fiber-optic sensor, a suitable signal modulation and demodulation method (such as the frequency-division multiplexing method, time-division multiplexing method, time-frequency-division multiplexing method, or coherence-division multiplexing method discussed in the previous chapter) and fiber-optic transmission network with a suitable topological pattern generally are required so that the information from different individual fiber-optic sensors can be separated.

The *distributed fiber-optic sensor* differs from the multiplexed one in that it consists of a continuous length of fiber, usually with no taps or branches along its length. A distributed fiber-optic sensor can detect not only the value of a parameter but also the location of the parameter. To create a distributed fiber-optic sensor, a suitable optical fiber that can modify the propagation of light reliably according to the parameter and a suitable method for absolute optical path difference measurement (such as optical time domain reflectometry (OTDR), white light interferometry, and optical FMCW interferometry) are generally required in order to determine both the value and the location of the parameter.

In the ideal case, distributed fiber-optic sensors should yield all spatial information along the whole length of the sensing fiber at a time. However, some distributed fiber-optic sensors, due to their restrictive nature, can give information about only one or a limited number of locations at one time, and these sensors are called *quasi-distributed fiber-optic sensors*.

Fiber-optic sensors are a broad topic. Systematic discussion of fiber-optic sensors is beyond the scope of this book. Therefore, we will concentrate on fiber-optic sensors based on optical FMCW interference.

Fiber-optic sensors based on the principle of optical interference are usually called *interferometric fiber-optic sensors*. Obviously, interferometric fiber-optic sensors are especially suitable for measuring displacement, distance, and optical refractive index because these quantities are directly related to the optical path difference in the interferometric fiber-optic sensor. However, due to the electro-optic effect, elasto-optic effect, or other physical and chemical effects, interferometric fiber-optic sensors can also be used to measure other physical quantities indirectly. Generally speaking, if any physical or chemical quantity can alter the optical path difference, it can be measured with an interferometric fiber-optic sensor.

In general, interferometric fiber-optic sensors have a higher level of accuracy than other types of fiber-optic sensors (such as intensity-based fiber-optic sensors) because interferometric fiber-optic sensors use an optical wavelength as the scale to measure the parameter. The FMCW interferometric fiber-optic sensors are superior to other types of interferometric fiber-optic sensors because they have no problems of dubious calibration and ambiguous fringe count and have the capability to measure the absolute value of the optical path difference.

In principle, the fiber-optic FMCW interferometers introduced in Chapter 7 and the multiplexed fiber-optic FMCW interferometers presented in Chapter 8 can be used as fiber-optic sensors to directly measure distance, displacement, and optical refractive index or indirectly measure other physical and chemical quantities. However, the configurations of those interferometers are generally not suited to practical situations. For instance, the reference fiber or the leading fiber of some interferometers in Chapter 7 or Chapter 8 may also be sensitive to the environment, so that the change of environmental conditions may make the interferometers unworkable. In the following sections, we will introduce some improved interferometric fiber-optic FMCW sensors that can overcome these problems and could be more suited to practical situations.

9.2 Fiber-optic FMCW Interferometric Displacement Sensors

Displacement measurement is an important topic in metrology. It is also essential to the measurement of other parameters. Fiber-optic FMCW displacement sensors can measure not only relative displacement but also absolute distance and speed of motion.

9.2.1 Reflectometric Single-Mode Fiber FMCW Displacement Sensors

Figure 9.1 shows a reflectometric single-mode fiber FMCW displacement sensor [53]. The sensor primarily consists of a Y-type single-mode fiber coupler (*FC*) stuck to a quarter-pitch gradient-index lens (*GL*) at the distant end of its output fiber. The outer planar surface of the gradient-index lens and the front surface of the object (*O*) to be measured are supposed to construct an air cavity.

An FMCW laser beam is coupled into one of the input fibers of *FC*, propagates through *GL*, and arrives on the front surface of *O*. A small fraction of the incident beam reflected from the outer planar surface of *GL* and a small fraction of the reflective beam reflected from the front surface of *O* are collected by the same gradient-index lens *GL* and used as the reference wave and signal wave, respectively. These two reflective beams mix coherently in the fiber to produce a beat signal. The beat signal propagates backward through the fiber coupler, emerges from the second input fiber of *FC*, and finally is received by a photodetector.

The quarter-pitch gradient-index lens used in this sensor has several functions: Firstly, it provides a reference beam with the reflection (Fresnel reflection) on its outer planar surface. Secondly, it collimates the exiting beam so

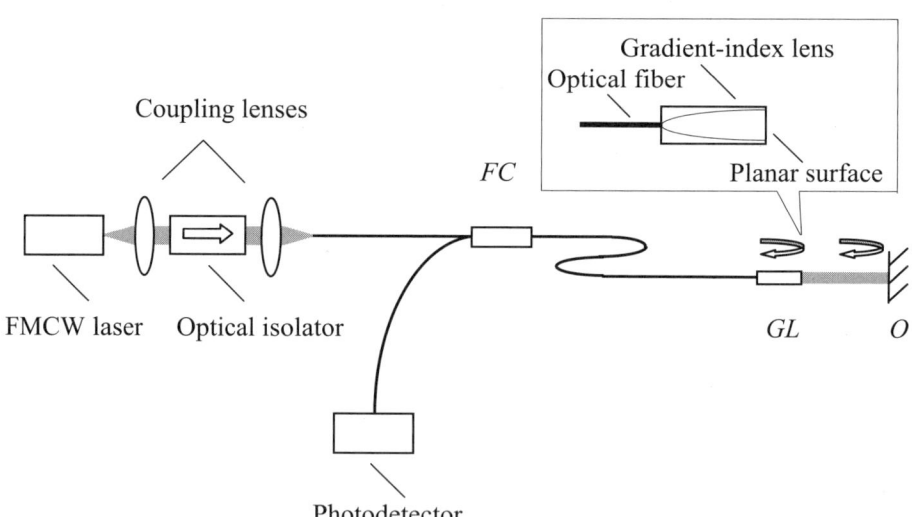

Figure 9.1. Reflectometric single-mode fiber FMCW displacement sensor. (After [53].)

that the exiting beam can propagate for a long distance without significant divergence. Thirdly, it efficiently collects the reflected reference and signal beams back to the fiber to produce a beat signal.

The optical path difference OPD between the two reflected beams can be written as

$$OPD = 2nd \, ,$$

(9-1)

where n is the refractive index of the air ($n \approx 1$), and d is the distance from the outer surface of the gradient-index lens to the front surface of the object to be measured. If the frequency of the laser is modulated with a sawtooth waveform, for instance, the intensity of the detected beat signal $I(t)$ in each modulation period can be written as

$$I(t) = I_0 \left[1 + V \cos\left(\frac{2\pi\Delta v v_m OPD}{c} t + \frac{2\pi}{\lambda_0} OPD \right) \right]$$

$$= I_0 [1 + V \cos(2\pi v_b t + \phi_{b0})]$$

(2-63)

where I_0 is the average intensity of the beat signal, V is the contrast of the beat signal, Δv is the optical frequency modulation excursion, v_m is the modulation frequency, c is the speed of light in free space, λ_0 is the central optical wavelength in free space, v_b and ϕ_{b0} are the frequency and initial phase of the beat signal, respectively.

Considering Equation (9-1), the frequency of the beat signal v_b can be written as

$$v_b = \frac{2nd\Delta v v_m}{c} \, ,$$

(9-2)

and the initial phase of the beat signal ϕ_{b0} can be written as

$$\phi_{b0} = \frac{4\pi nd}{\lambda_0} \, .$$

(9-3)

Therefore, measuring the frequency of the beat signal, we can obtain the absolute distance d between the gradient-index lens and the object,

$$d = \frac{c}{2n\Delta v v_m} v_b \, ,$$

(9-4)

and measuring the change of the initial phase of the beat signal, we can obtain the relative displacement Δd of the object,

$$\Delta d = \frac{\lambda_0}{4\pi n} \Delta\phi_{b0} .$$

(9-5)

Moreover, if the frequency of the laser is modulated with a triangular waveform, according to the conclusion in Section 2.2.2, we can find the average Doppler frequency shift $\overline{v_D}$ by

$$\overline{v_D} = \frac{1}{2}\left(\overline{v_{br}}' - \left|\overline{v_{bf}}'\right|\right) ,$$

(9-6)

where $\overline{v_{br}}'$ is the average beat frequency in the rising periods and $\left|\overline{v_{bf}}'\right|$ is the average beat frequency in the falling periods. Consequently, we can obtain the speed s of the moving object by

$$s = \frac{\lambda_0}{2n}\overline{v_D} .$$

(9-7)

The measurement accuracy of this fiber-optic displacement sensor is mainly determined by the phase noise and frequency drift of the laser. The phase noise associated with the laser makes the phase of the beat signal unstable, which, in turn, directly affects the phase measurement accuracy and indirectly increases the uncertainty in the frequency measurement. The frequency drift of the laser is the major problem for long-term stability, but this effect can be minimized by using the methods introduced in Section 3.3.7. The nonlinear response of the frequency modulation of the laser also makes the beat frequency measurement imprecise; this problem, however, can be minimized by using an averaging method. Due to the nature of the reflectometric structure, the measurement range of this sensor is limited to half of the coherence length of the laser.

The advantage of this fiber-optic displacement sensor is that it is immune to environment interference due to the nature of the single-arm Fizeau interferometer structure. In other words, because the two reflected beams in the sensor always propagate within the same length of single-mode fiber except in the air gap between the gradient-index lens and the object, the effects of the environmental conditions (such as the change of temperature or stretch of fiber) on the two interfering beams are naturally compensated. Therefore, this sensor is very stable and can have a long length of leading fiber to detect a distant target. In addition, the small gradient-index lens (generally 2 millimeters in diameter and

5 millimeters in length) and the flexible leading fiber make the sensor workable in some complicated situations.

9.2.2 Multiplexed Reflectometric Single-Mode Fiber FMCW Displacement Sensors

The reflectometric single-mode fiber FMCW displacement sensor discussed previously can be modified into a multiplexed displacement sensor by using the frequency-division multiplexing method. Figure 9.2 shows a double-sensor multiplexed reflectometric single-mode fiber FMCW displacement sensor, in which an X-type fiber coupler (FC) is employed [54]. Each output fiber of the X-type fiber coupler is stuck to a quarter-pitch gradient-index lens at the end as a probe. The two gradient-index lenses (GL_1 and GL_2) are placed at different distances from two objects (O_1 and O_2) so that the two air cavities are different in length and thus the two beat signals are different in frequency. These two beat signals are detected by the same photodetector but are finally separated by using two electric band-pass filters.

Note that, in order to keep the two beat signals from interfering coherently, the two leading fibers (i.e., the two output fibers of FC) should be different in length, and the length difference (represented by a single circle in Figure 9.2) should exceed half the coherence length of the laser.

In the case of sawtooth-wave modulation, for instance, the intensity of the detected signal in any modulation period $I'(t)$ can be written as

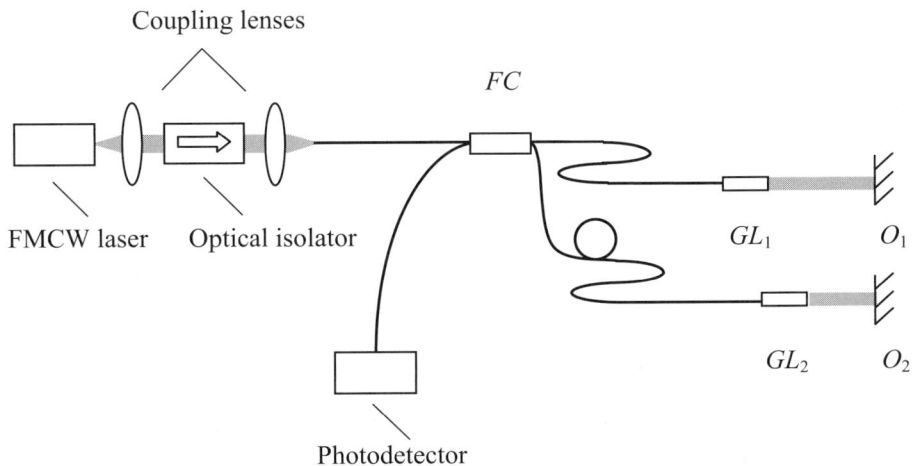

Figure 9.2. Double-sensor multiplexed reflectometric single-mode fiber FMCW displacement sensor. (After [54].)

$$I'(t) = (I_{01} + I_{02}) + \left[I_{01}V_1 \cos\left(\frac{2\pi\Delta\nu\nu_m OPD_1}{c} t + \frac{2\pi}{\lambda_0} OPD_1 \right) \right.$$

$$+ I_{02}V_2 \cos\left(\frac{2\pi\Delta\nu\nu_m OPD_2}{c} t + \frac{2\pi}{\lambda_0} OPD_2 \right) \right]$$

$$= (I_{01} + I_{02}) + [I_{01}V_1 \cos(2\pi\nu_{b1}t + \phi_{b01}) + I_{02}V_2 \cos(2\pi\nu_{b2}t + \phi_{b02})], \quad (9\text{-}8)$$

where I_{01}, V_1, ν_{b1}, and ϕ_{b01} are the average intensity, contrast, frequency, and initial phase of the beat signal from the first sensor, respectively; I_{02}, V_2, ν_{b2}, and ϕ_{b02} are the average intensity, contrast, frequency, and initial phase of the beat signal from the second sensor, respectively; $\Delta\nu$ is the optical frequency modulation excursion, ν_m is the modulation frequency, c is the speed of light in free space, λ_0 is the central optical wavelength in free space.

The initial phases ϕ_{b01} and ϕ_{b02} are given by

$$\phi_{b01} = \frac{4\pi n d_1}{\lambda_0}, \quad (9\text{-}9)$$

$$\phi_{b02} = \frac{4\pi n d_2}{\lambda_0}, \quad (9\text{-}10)$$

where n is the refractive index of the air ($n \approx 1$), d_1 is the distance from GL_1 to O_1, and d_2 is the distance from GL_2 to O_2. Obviously, the displacements of the two moving objects Δd_1 and Δd_2 can be determined by

$$\Delta d_1 = \frac{\lambda_0}{4\pi n} \Delta\phi_{b01}, \quad (9\text{-}11)$$

$$\Delta d_2 = \frac{\lambda_0}{4\pi n} \Delta\phi_{b02}. \quad (9\text{-}12)$$

The accuracy of the multiplexed displacement sensor mainly depends on the cross talk between the two beat signals. In the entire time domain, the intensity of the beat signal from each individual displacement sensor $I(t)$ can be written as

$$I(t) = I_0 \{1 + [V\cos(\alpha\tau t + \omega_0\tau)\operatorname{rect}_{T_m}(t) \otimes \sum_{m=-\infty}^{\infty} \delta(t - mT_m)]\}$$

$$= I_0 \{1 + [V\cos(\omega_b t + \phi_{b0})\operatorname{rect}_{T_m}(t) \otimes \sum_{m=-\infty}^{\infty} \delta(t - mT_m)]\} \qquad (2\text{-}61)$$

where I_0, V, ω_b, and ϕ_{b0} are the average intensity, contrast, angular frequency, and initial phase of the individual beat signal, respectively, and T_m is the period of the modulation signal. The Fourier spectrum of this signal $I(\omega)$ can be written as

$$I(\omega) = I_0 \left(2\pi\delta(\omega) + \pi V \left\{ \frac{\sin\left[\dfrac{(\omega+\omega_b)T_m}{2}\right]}{\dfrac{(\omega+\omega_b)T_m}{2}} e^{-j\phi_{b0}} + \frac{\sin\left[\dfrac{(\omega-\omega_b)T_m}{2}\right]}{\dfrac{(\omega-\omega_b)T_m}{2}} e^{j\phi_{b0}} \right\} \right.$$

$$\left. \sum_{m=-\infty}^{\infty} \delta(\omega - m\omega_m) \right) \qquad ,(9\text{-}13)$$

where ω_m is the modulation angular frequency ($\omega_m = 2\pi/T_m$), or

$$I(\omega) = I_0 \left(2\pi\delta(\omega) + \pi V \left\{ \operatorname{Sinc}\left[\frac{(\omega+\omega_b)T_m}{2}\right] e^{-j\phi_{b0}} + \operatorname{Sinc}\left[\frac{(\omega-\omega_b)T_m}{2}\right] e^{j\phi_{b0}} \right\} \right.$$

$$\left. \sum_{m=-\infty}^{\infty} \delta(\omega - m\omega_m) \right) \qquad (9\text{-}14)$$

Equation (9-14) shows that the spectrum of the beat signal is a series of δ functions enveloped by a Sinc function. The maximum value of the Sinc function is located at ω_b, and the zero-point period of the Sinc function equals ω_m, as shown in Figure 9.3. Each δ function gives a harmonic component, while each harmonic component contains the information of the initial phase ϕ_{b0}. If we use an electric band-pass filter to choose any harmonic component $m\omega_m$, the filtered signal $i_m(t)$ will be

$$i_m(t) = A_m \cos(m\omega_m t + \phi_{b0}), \qquad (9\text{-}15)$$

where A_m is a constant representing the intensity of the mth harmonic component, given by

$$A_m = I_0 V \operatorname{Sinc}\left[\frac{(m\omega_m - \omega_b)T_m}{2}\right].$$

(9-16)

Generally, we choose the most intensive harmonic component $M\omega_m$ ($M\omega_m \approx \omega_b$, where M is the order of the most intensive harmonic component), so that the output signal from the electric filter would be

$$i_M(t) = A_M \cos(M\omega_m t + \phi_{b0}).$$

(9-17)

If the two sensors are multiplexed, the intensity of the combined signal $I'(t)$ can be written as

$$I'(t) = (I_{01} + I_{02}) + [I_{01}V_1 \cos(\omega_{b1}t + \phi_{b01}) + I_{02}V_2 \cos(\omega_{b2}t + \phi_{b02})]\operatorname{rect}_{T_m}(t)$$

$$\otimes \sum_{m=-\infty}^{\infty} \delta(t - mT_m)$$

,

(9-18)

where I_{01}, V_1, ω_{b1}, and ϕ_{b01} are the average intensity, contrast, angular frequency, and initial phase of the beat signal from the first sensor, respectively; I_{02}, V_2, ω_{b2}, and ϕ_{b02} are the average intensity, contrast, angular frequency, and initial phase of the beat signal from the second sensor, respectively. The Fourier spectrum of the combined signal will be

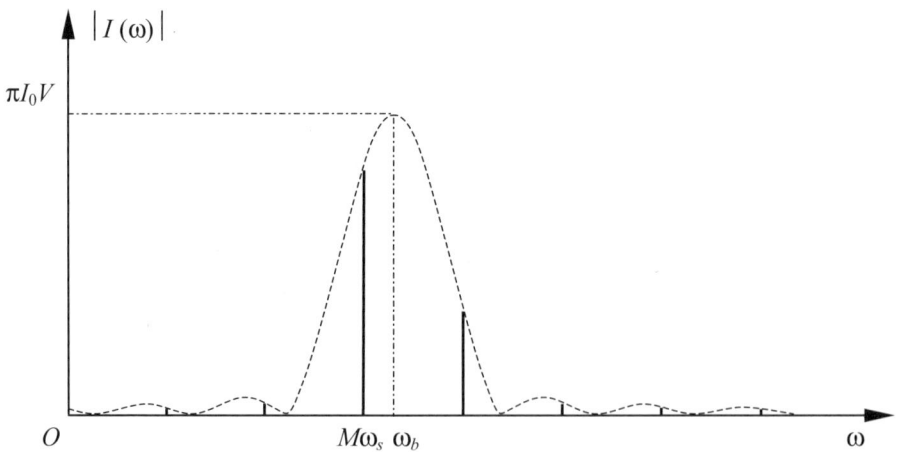

Figure 9.3. Spectrum of the signal from a single-mode fiber FMCW displacement sensor.

$$I'(\omega) = 2\pi(I_{01} + I_{02})\delta(\omega) + \pi\left\{I_{01}V_1 \frac{\sin\left[\dfrac{(\omega+\omega_{b1})T_m}{2}\right]}{\dfrac{(\omega+\omega_{b1})T_m}{2}} e^{-j\phi_{b01}} + I_{02}V_2 \frac{\sin\left[\dfrac{(\omega+\omega_{b2})T_m}{2}\right]}{\dfrac{(\omega+\omega_{b2})T_m}{2}} e^{-j\phi_{b02}}\right.$$

$$\left. + I_{01}V_1 \frac{\sin\left[\dfrac{(\omega-\omega_{b1})T_m}{2}\right]}{\dfrac{(\omega-\omega_{b1})T_m}{2}} e^{j\phi_{b01}} + I_{02}V_2 \frac{\sin\left[\dfrac{(\omega-\omega_{b2})T_m}{2}\right]}{\dfrac{(\omega-\omega_{b2})T_m}{2}} e^{j\phi_{b02}}\right\} \sum_{m=-\infty}^{\infty}\delta(\omega-m\omega_m)$$

$$(9\text{-}19)$$

or

$$I'(\omega) = 2\pi(I_{01} + I_{02})\delta(\omega) + \pi\left\{I_{01}V_1 \, \text{Sinc}\left[\frac{(\omega+\omega_{b1})T_m}{2}\right] e^{-j\phi_{b01}}\right.$$

$$+ I_{02}V_2 \, \text{Sinc}\left[\frac{(\omega+\omega_{b2})T_m}{2}\right] e^{-j\phi_{b02}} + I_{01}V_1 \, \text{Sinc}\left[\frac{(\omega-\omega_{b1})T_m}{2}\right] e^{j\phi_{b01}}$$

$$\left. + I_{02}V_2 \, \text{Sinc}\left[\frac{(\omega-\omega_{b2})T_m}{2}\right] e^{j\phi_{b02}}\right\} \sum_{m=-\infty}^{\infty}\delta(\omega-m\omega_m)$$

$$(9\text{-}20)$$

Obviously, the signal spectrum of the double-sensor multiplexed FMCW displacement sensor is still a series of δ functions, but they are enveloped by the sum of two Sinc functions whose maximum values are located at ω_{b1} and ω_{b2}, respectively, and whose zero-point periods are ω_m. In other words, the spectrum of the combined signal is the superposition of the spectra of two individual beat signals, as shown in Figure 9.4, where the solid line represents the spectrum of the beat signal from the first sensor, the dashed line represents the spectrum of the beat signal from the second sensor (for easy understanding, the two spectra have not been added; instead, one of them has been shifted slightly). Therefore, each harmonic component of the combined signal actually consists of two parts contributed by the two different sensors. If we choose any harmonic component to retrieve the phase information detected by a specific sensor, one part is the effective signal, while another part will be the cross talk.

In general, the frequencies of the two beat signals are arranged close to different specified harmonic frequencies (e.g., $\omega_{b1} \approx M\omega_m$, $\omega_{b2} \approx N\omega_m$, $N \neq M$). If the Mth harmonic is selected for determining the displacement of the first object, the output signal from the electric band-pass filter $i_M'(t)$ will be

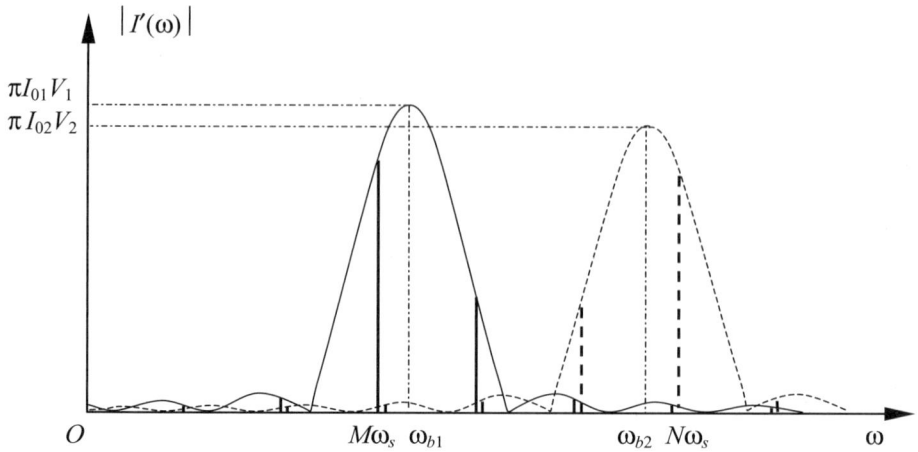

Figure 9.4. Spectrum of the signal from a double-sensor multiplexed single-mode fiber FMCW displacement sensor.

$$i'_M(t) = B_M \cos\left\{ M\omega_m t + \arg\left\{ I_{01}V_1 \text{Sinc}\left[\frac{(M\omega_m - \omega_{b1})T_m}{2}\right]e^{j\phi_{b01}} \right.\right.$$
$$\left.\left. + I_{02}V_2 \text{Sinc}\left[\frac{(M\omega_m - \omega_{b2})T_m}{2}\right]e^{j\phi_{b02}} \right\}\right\}$$

, (9-21)

where B_M is a constant given by

$$B_M = \left| I_{01}V_1 \text{Sinc}\left[\frac{(M\omega_m - \omega_{b1})T_m}{2}\right]e^{j\phi_{b01}} + I_{02}V_2 \text{Sinc}\left[\frac{(M\omega_m - \omega_{b2})T_m}{2}\right]e^{j\phi_{b02}} \right|$$

(9-22)

Comparing Equation (9-21) with Equation (9-17), the phase error in the first sensor due to the cross talk from the second sensor will be

$$\delta\phi_{b01} = \arg\left\{ I_{01}V_1 \text{Sinc}\left[\frac{(M\omega_m - \omega_{b1})T_m}{2}\right]e^{j\phi_{b01}} + I_{02}V_2 \text{Sinc}\left[\frac{(M\omega_m - \omega_{b2})T_m}{2}\right]e^{j\phi_{b02}} \right\}$$
$$- \phi_{b01}$$

(9-23)

Similarly, it can be derived that, if the Nth harmonic is selected for determining the displacement of the second object, the phase error in the second sensor due to the cross talk from the first sensor will be

$$\delta\phi_{b02} = \arg\left\{ I_{01}V_1 \text{Sinc}\left[\frac{(N\omega_m - \omega_{b1})T_m}{2}\right]e^{j\phi_{b01}} + I_{02}V_2 \text{Sinc}\left[\frac{(N\omega_m - \omega_{b2})T_m}{2}\right]e^{j\phi_{b02}} \right\}$$
$$- \phi_{b02}$$

(9-24)

The phase error due to cross talk can also be calculated by using a phasor diagram. Phasors are rotating vectors that are frequently used to express harmonic vibrations or waves. In the case of the double-sensor multiplexed fiber-optic displacement sensor, for instance, we can use the phasor $\boldsymbol{P}_{1M}$ to represent the Mth harmonic component of the beat signal from the first sensor and use the phasor $\boldsymbol{P}_{2M}$ to represent the Mth harmonic component of the beat signal from the second sensor,

$$\boldsymbol{P}_{1M} = I_{01}V_1 \text{Sinc}\left[\frac{(M\omega_m - \omega_{b1})T_m}{2}\right]e^{j\phi_{b01}}$$

(9-25)

$$\boldsymbol{P}_{2M} = I_{02}V_2 \text{Sinc}\left[\frac{(M\omega_m - \omega_{b2})T_m}{2}\right]e^{j\phi_{b02}}$$

(9-26)

If we choose the Mth harmonic component of the combined signal to determine the phase detected by the first sensor (assuming $\omega_{b1} \approx M\omega_m$), then $\boldsymbol{P}_{1M}$ is the effective signal, but $\boldsymbol{P}_{2M}$ is the cross talk. The phasor $\boldsymbol{P}_M$ of the Mth harmonic component of the combined signal could be

$$\boldsymbol{P}_M = \boldsymbol{P}_{1M} + \boldsymbol{P}_{2M}$$
$$= I_{01}V_1 \text{Sinc}\left[\frac{(M\omega_m - \omega_{b1})T_m}{2}\right]e^{j\phi_{b01}} + I_{02}V_2 \text{Sinc}\left[\frac{(M\omega_m - \omega_{b2})T_m}{2}\right]e^{j\phi_{b02}}$$

(9-27)

Figure 9.5(a) shows the relationship of $\boldsymbol{P}_{1M}$, $\boldsymbol{P}_{2M}$, and $\boldsymbol{P}_M$. Obviously, the angle between $\boldsymbol{P}_{1M}$ and $\boldsymbol{P}_M$ equals the phase error $\delta\phi_{b01}$ due to cross talk. Particularly when $\boldsymbol{P}_M \perp \boldsymbol{P}_{2M}$ (i.e., $\phi_{b02} \approx \phi_{b01}\pm\pi/2$), $\delta\phi_{b01}$ reaches an extreme $(\delta\phi_{b01})_{extr}$

$$(\delta\phi_{b01})_{extr} = \arcsin(\frac{P_{2M}}{P_{1M}})$$

$$= \arcsin\left\{\frac{I_{02}V_2 \ \text{Sinc}\left[\dfrac{(M\omega_m - \omega_{b2})T_m}{2}\right]}{I_{01}V_1 \ \text{Sinc}\left[\dfrac{(M\omega_m - \omega_{b1})T_m}{2}\right]}\right\},$$

(9-28)

where P_{1M} and P_{2M} are the amplitudes of the phasors $\boldsymbol{P}_{1M}$ and $\boldsymbol{P}_{2M}$, respectively, as shown in Figure 9.5(b). In practice, ω_{b1} is close to $M\omega_m$, but ω_{b2} is far away from $M\omega_m$, thus, $P_{1M} \gg P_{2M}$, and therefore Equation (9-28) can be simplified as

$$(\delta\phi_{b01})_{extr} \approx \frac{P_{2M}}{P_{1M}}$$

$$= \frac{I_{01}V_2 \ \text{Sinc}\left[\dfrac{(M\omega_m - \omega_{b2})T_m}{2}\right]}{I_{02}V_1 \ \text{Sinc}\left[\dfrac{(M\omega_m - \omega_{b1})T_m}{2}\right]}.$$

(9-29)

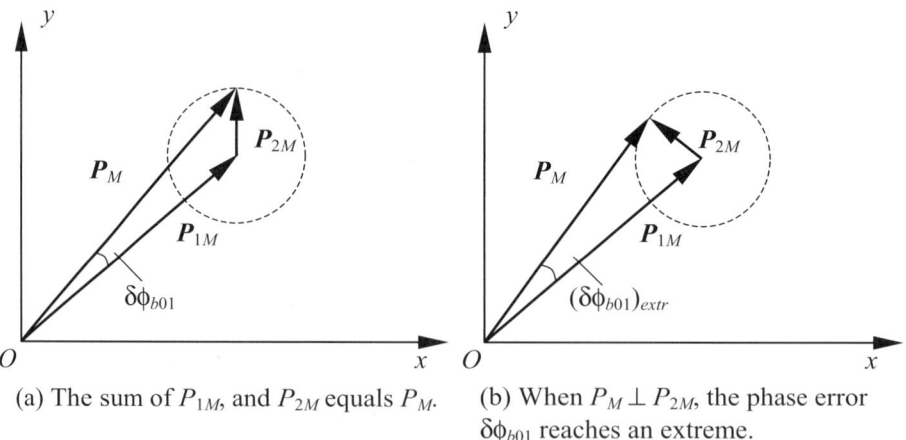

(a) The sum of P_{1M}, and P_{2M} equals P_M.   (b) When $P_M \perp P_{2M}$, the phase error $\delta\phi_{b01}$ reaches an extreme.

Figure 9.5. Phasor diagram for calculation of the cross talk in the double-sensor multiplexed single-mode fiber FMCW displacement sensor.

The maximum phase error $(\delta\phi_{b01})_{max}$ is related to the maximum deviations of the two beat frequencies from the specified harmonic frequencies and can be written as

$$(\delta\phi_{b01})_{max} = \arcsin\left\{\frac{I_{02}V_2\,\mathrm{Sinc}\left[\dfrac{(M\omega_m - \Omega_{b2})T_m}{2}\right]}{I_{01}V_1\,\mathrm{Sinc}\left[\dfrac{(M\omega_m - \Omega_{b1})T_m}{2}\right]}\right\}$$

$$\approx \frac{I_{02}V_2\,\mathrm{Sinc}\left[\dfrac{(M\omega_m - \Omega_{b2})T_m}{2}\right]}{I_{01}V_1\,\mathrm{Sinc}\left[\dfrac{(M\omega_m - \Omega_{b1})T_m}{2}\right]} \quad , \tag{9-30}$$

where Ω_{b1} and Ω_{b2} are the maximum angular frequencies of the two beat signals. The maximum displacement error $(\delta\Delta d_1)_{max}$ can be written as

$$(\delta\Delta d_1)_{max} = \frac{\lambda_0}{4\pi n}\arcsin\left\{\frac{I_{02}V_2\,\mathrm{Sinc}\left[\dfrac{(M\omega_m - \Omega_{b2})T_m}{2}\right]}{I_{01}V_1\,\mathrm{Sinc}\left[\dfrac{(M\omega_m - \Omega_{b1})T_m}{2}\right]}\right\}$$

$$\approx \frac{\lambda_0 I_{02}V_2\,\mathrm{Sinc}\left[\dfrac{(M\omega_m - \Omega_{b2})T_m}{2}\right]}{4\pi n I_{01}V_1\,\mathrm{Sinc}\left[\dfrac{(M\omega_m - \Omega_{b1})T_m}{2}\right]} \quad , \tag{9-31}$$

where λ_0 is the central optical wavelength in free space.

In the best case, if the dynamic range of the displacement sensor is especially narrow (i.e., $\Omega_{b1} \approx M\omega_m$, $\Omega_{b2} \approx N\omega_m$), there will be nearly no cross talk $((\delta\phi_{b1})_{max} \approx 0$, $(\delta\Delta d_1)_{max} \approx 0)$. Under this situation, the power of each individual beat signal concentrates on a single spectral line at different specified harmonic positions.

In the worst case, for instance, assuming the measurement range of the displacement sensor is equal to the modulation frequency ω_m (i.e., $\Omega_{b1} = M\omega_m + \omega_m/2$, $\Omega_{b2} = N\omega_m + \omega_m/2$), the maximum phase error $(\delta\phi_{b01})_{max}$ due to signal cross talk will be

$$(\delta\phi_{b01})_{max} = \frac{I_{02}V_2}{I_{01}V_1[2(N-M)+1]} \quad , \tag{9-32}$$

Table 9.1. The maximum phase and displacement errors in the double-sensor multiplexed single-mode fiber FMCW displacement sensor. (Assuming $\Omega_{b1} = M\omega_m + \omega_m/2$, $\Omega_{b2} = N\omega_m + \omega_m/2$, $I_{01}V_1 = I_{02}V_2$, $n = 1$.)

$N-M$ (ω_m)	1	2	3	4	5
$(\delta\phi_{b01})_{max}$ (radians)	1/3	1/5	1/7	1/9	1/11
$(\delta\Delta d_1)_{max}$ (λ_0)	$1/12\pi$	$1/20\pi$	$1/28\pi$	$1/36\pi$	$1/44\pi$

and the corresponding maximum displacement error $(\delta\Delta d_1)_{max}$ will be

$$(\delta\Delta d_1)_{max} = \frac{I_{02}V_2\lambda_0}{4\pi n I_{01}V_1[2(N-M)+1]}. \tag{9-33}$$

Obviously, $(\delta\phi_{b01})_{max}$ and $(\delta\Delta d_1)_{max}$ are related to the ratios of the average intensity and contrast of the two beat signals and inversely proportional to the gap of the two beat frequencies $(N-M)$. Increasing the gap of the beat frequencies can reduce the cross talk, but it requires one of the sensors to have a longer air cavity. This will reduce the intensity of the reflected beam and increase the coherence noise. In general, the frequency gap between the two beat signals is selected from 2 to 4. Table 9.1 shows the maximum phase errors and the maximum displacement errors corresponding to different frequency gaps.

It should be noted that, in practice, other factors, such as nonlinear frequency modulation response, intensity modulation, and phase noise of the laser, cause the energy of the beat signal from each individual interferometer to spread over a wide range rather than to concentrate on a single spectral line. Therefore, the real cross talk may be bigger than the theoretical prediction. Moreover, since different sensors have different lengths of air cavities and thus different signal contrasts, different sensors generally have different amounts of cross talk. The longer the cavity, the bigger the cross talk will be.

Figure 9.6 shows the scheme of a triple-sensor multiplexed reflectometric single-mode fiber FMCW displacement sensor, which employs a 2×3 multiple-port single-mode fiber coupler (FC) and three quarter-pitch gradient-index lenses (GL_1, GL_2 and GL_3) to measure three objects (O_1, O_2 and O_3). Figure 9.7 demonstrates another version of the triple-sensor multiplexed fiber-optic FMCW displacement sensor, which employs two X-type single-mode fiber couplers (FC_1 and FC_2).

The intensity of the combined signal $I''(t)$ from the triple-sensor multiplexed displacement sensor can be written as

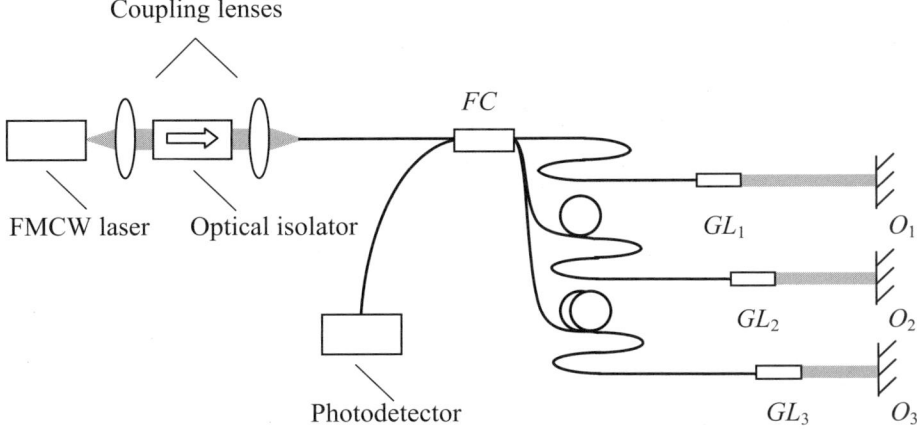

Figure 9.6. Triple-sensor multiplexed single-mode fiber FMCW displacement sensor.

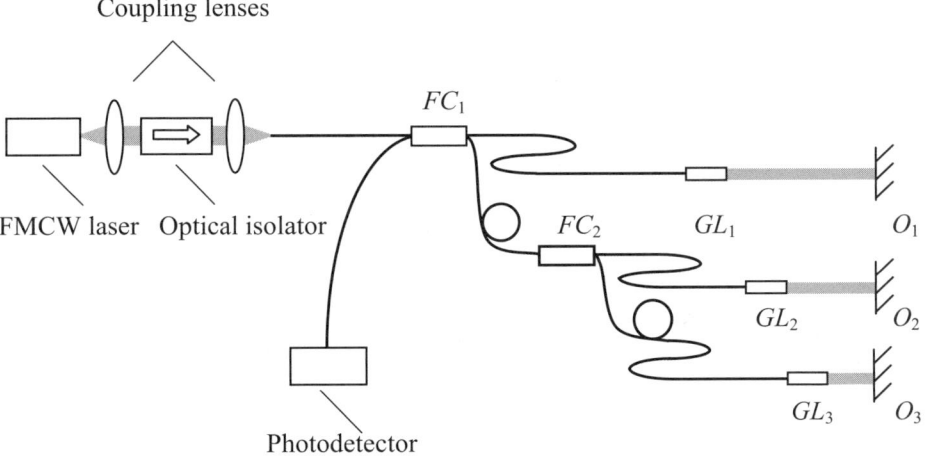

Figure 9.7. A different version of the triple-sensor multiplexed single-mode fiber FMCW displacement sensor.

$$I''(t) = (I_{01} + I_{02} + I_{03}) + [I_{01}V_1 \cos(\omega_{b1}t + \phi_{b01}) + I_{02}V_2 \cos(\omega_{b2}t + \phi_{b02})$$

$$+ I_{03}V_3 \cos(\omega_{b3}t + \phi_{b03})]\,\mathrm{rect}_{T_m}(t) \otimes \sum_{m=-\infty}^{\infty} \delta(t - mT_m)$$

$$(9\text{-}34)$$

where I_{01}, V_1, ω_{b1}, and ϕ_{b01} are the average intensity, contrast, angular frequency, and initial phase of the beat signal from the first sensor, respectively; I_{02}, V_2, ω_{b2}, and ϕ_{b02} are the average intensity, contrast, angular frequency, and initial phase of the beat signal from the second sensor, respectively; I_{03}, V_3, ω_{b3} and ϕ_{b03} are the average intensity, contrast, angular frequency and initial phase of the beat signal from the third sensor respectively. The Fourier spectrum of the combined signal would be

$$
\begin{aligned}
I''(\omega) = 2\pi(I_{01} + I_{02} + I_{03})\delta(\omega) + 2\pi^2 \Bigg\{ &I_{01}V_1 \frac{\sin\left[\frac{(\omega+\omega_{b1})T_m}{2}\right]}{\frac{(\omega+\omega_{b1})T_m}{2}} e^{-j\phi_{b01}} \\
+ I_{02}V_2 \frac{\sin\left[\frac{(\omega+\omega_{b2})T_m}{2}\right]}{\frac{(\omega+\omega_{b2})T_m}{2}} e^{-j\phi_{b02}} &+ I_{03}V_3 \frac{\sin\left[\frac{(\omega+\omega_{b3})T_m}{2}\right]}{\frac{(\omega+\omega_{b3})T_m}{2}} e^{-j\phi_{b03}} \\
+ I_{01}V_1 \frac{\sin\left[\frac{(\omega-\omega_{b1})T_m}{2}\right]}{\frac{(\omega-\omega_{b1})T_m}{2}} e^{j\phi_{b01}} &+ I_{02}V_2 \frac{\sin\left[\frac{(\omega-\omega_{b2})T_m}{2}\right]}{\frac{(\omega-\omega_{b2})T_m}{2}} e^{j\phi_{b02}} \\
+ I_{03}V_3 \frac{\sin\left[\frac{(\omega-\omega_{b3})T_m}{2}\right]}{\frac{(\omega-\omega_{b3})T_m}{2}} e^{j\phi_{b03}} &\Bigg\} \sum_{m=-\infty}^{\infty} \delta(\omega - m\omega_m)
\end{aligned}
\qquad (9\text{-}35)
$$

or

$$
\begin{aligned}
I''(\omega) = 2\pi(I_{01} + I_{02} + I_{03})\delta(\omega) + +2\pi^2 \Bigg\{ &I_{01}V_1 \operatorname{Sinc}\left[\frac{(\omega+\omega_{b1})T_m}{2}\right] e^{-j\phi_{b01}} \\
+ I_{02}V_2 \operatorname{Sinc}\left[\frac{(\omega+\omega_{b2})T_m}{2}\right] e^{-j\phi_{b02}} &+ I_{03}V_3 \operatorname{Sinc}\left[\frac{(\omega+\omega_{b3})T_m}{2}\right] e^{-j\phi_{b03}} \\
+ I_{01}V_1 \operatorname{Sinc}\left[\frac{(\omega-\omega_{b1})T_m}{2}\right] e^{j\phi_{b01}} &+ I_{02}V_2 \operatorname{Sinc}\left[\frac{(\omega-\omega_{b2})T_m}{2}\right] e^{j\phi_{b02}} \\
+ I_{03}V_3 \operatorname{Sinc}\left[\frac{(\omega-\omega_{b3})T_m}{2}\right] e^{j\phi_{b03}} &\Bigg\} \sum_{m=-\infty}^{\infty} \delta(\omega - m\omega_m)
\end{aligned}
\qquad (9\text{-}36)
$$

Similarly, the signal spectrum of the triple-sensor multiplexed FMCW displacement sensor is also a series of δ functions, but they are enveloped by the sum of three Sinc functions, whose maximum values are located at ω_{b1}, ω_{b2} and ω_{b3} respectively, and whose zero-point periods are ω_m. In other words, the spectrum of the combined signal is the superposition of the spectrums of three individual beat signals, as shown in Figure 9.8, where the solid line represents the spectrum of the beat signal from the first sensor, the dashed line represents the spectrum of the beat signal from the second sensor, and the dot-dashed line represents the spectrum of the beat signal from the third sensor (for clarity, the spectra have not been added but shifted). Therefore, each harmonic component of the combined signal actually consists of three parts contributed by the three different sensors. If we choose any harmonic component to retrieve the phase information detected by a specific sensor, one part is the effective signal, while another two parts are the cross talk.

If the Mth harmonic component is picked out for determining the displacement detected by the first sensor (assuming $\omega_{b1} \approx M\omega_m$), the output signal from the electric band-pass filter will be

$$i''_M(t) = C_M \cos\left(M\omega_m t + \arg\left\{ I_{01} V_1 \operatorname{Sinc}\left[\frac{(M\omega_m - \omega_{b1})T_m}{2} \right] e^{j\phi_{b01}} \right.\right.$$
$$\left.\left. + I_{02} V_2 \operatorname{Sinc}\left[\frac{(M\omega_m - \omega_{b2})T_m}{2} \right] e^{j\phi_{b02}} + I_{03} V_3 \operatorname{Sinc}\left[\frac{(M\omega_m - \omega_{b3})T_m}{2} \right] e^{j\phi_{b03}} \right\}\right)$$

$$(9\text{-}37)$$

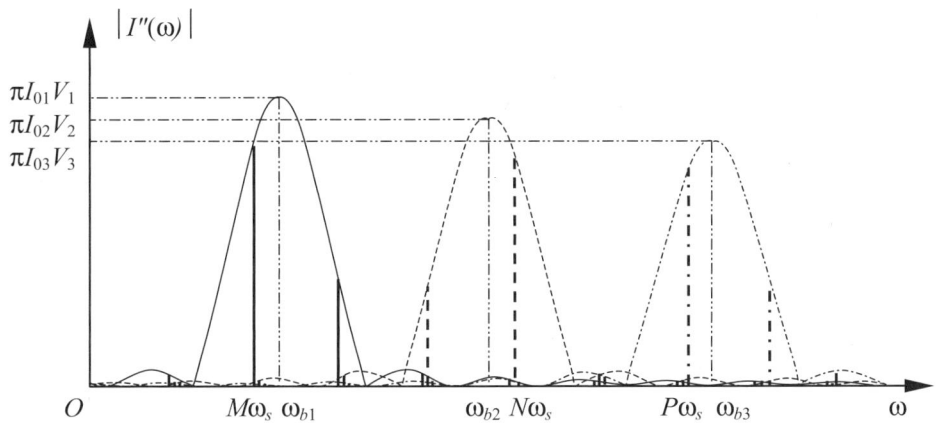

Figure 9.8. Spectrum of the signal from a triple-sensor frequency-division multiplexed single-mode fiber FMCW displacement sensor.

where C_M is a constant given by

$$C_M = \left| I_{01}V_1 \,\mathrm{Sinc}\left[\frac{(M\omega_m - \omega_{b1})T_m}{2}\right] e^{j\phi_{b01}} + I_{02}V_2 \,\mathrm{Sinc}\left[\frac{(M\omega_m - \omega_{b2})T_m}{2}\right] e^{j\phi_{b02}} \right.$$
$$\left. + I_{03}V_3 \,\mathrm{Sinc}\left[\frac{(M\omega_m - \omega_{b3})T_m}{2}\right] e^{j\phi_{b03}} \right|$$

$$(9\text{-}38)$$

The phase error due to the cross talk from the two other sensors would be

$$\delta\phi_{b01} = \arg\left\{ I_{01}V_1 \,\mathrm{Sinc}\left[\frac{(M\omega_m - \omega_{b1})T_m}{2}\right] e^{j\phi_{b01}} + I_{02}V_2 \,\mathrm{Sinc}\left[\frac{(M\omega_m - \omega_{b2})T_m}{2}\right] e^{j\phi_{b02}} \right.$$
$$\left. + I_{03}V_3 \,\mathrm{Sinc}\left[\frac{(M\omega_m - \omega_{b3})T_m}{2}\right] e^{j\phi_{b03}} \right\} - \phi_{b01}$$

$$(9\text{-}39)$$

The phasor diagram of the filtered signal is shown in Figure 9.9(a), where P_{1M} represents the Mth harmonic component of the beat signal from the first sensor, P_{2M} represents the Mth harmonic component of the beat signal from the second sensor, P_{3M} represents the Mth harmonic component of the beat signal from the third sensor, and P_M represents the Mth harmonic component of the combined signal. These phasors are given by

$$P_{1M} = I_{01}V_1 \,\mathrm{Sinc}\left[\frac{(M\omega_m - \omega_{b1})T_m}{2}\right] e^{j\phi_{b01}} \qquad (9\text{-}40)$$

$$P_{2M} = I_{02}V_2 \,\mathrm{Sinc}\left[\frac{(M\omega_m - \omega_{b2})T_m}{2}\right] e^{j\phi_{b02}} \qquad (9\text{-}41)$$

$$P_{3M} = I_{03}V_3 \,\mathrm{Sinc}\left[\frac{(M\omega_m - \omega_{b3})T_m}{2}\right] e^{j\phi_{b03}} \qquad (9\text{-}42)$$

$$P_M = P_{1M} + P_{2M} + P_{3M} \qquad (9\text{-}43)$$

Obviously, the angle between P_{1M} and P_M is equal to the phase error $\delta\phi_{b01}$. Particularly when $P_M \perp P_{2M}$ and $P_{2M} \parallel P_{3M}$, the phase error $\delta\phi_{b01}$ reaches an

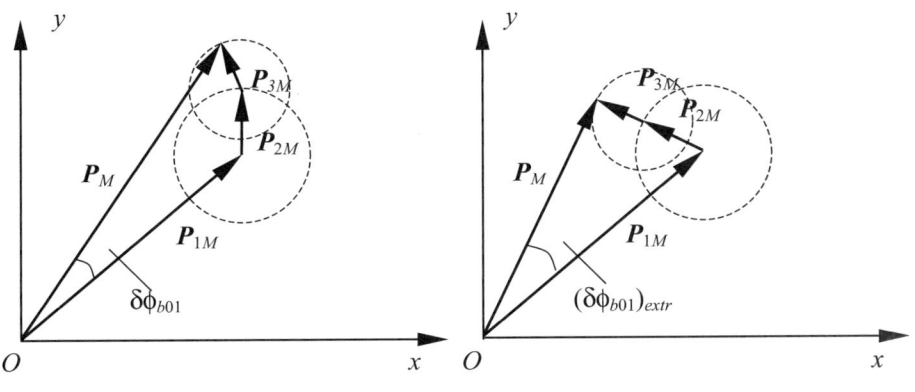

(a) The sum of P_{1M}, P_{2M} and P_{3M} equals P_M.

(b) When $P_M \perp P_{2M}$ and $P_{2M} \parallel P_{3M}$, the phase error $\delta\phi_{b01}$ reaches an extreme.

Figure 9.9. Phasor diagram for calculation of the cross talk in the triple-sensor frequency-division multiplexed single-mode fiber FMCW displacement sensor.

extreme $(\delta\phi_{b01})_{extr}$,

$$
\begin{aligned}
(\delta\phi_{b01})_{extr} &= \arcsin\left(\frac{P_{2M}+P_{3M}}{P_{1M}}\right) \\
&= \arcsin\left\{\frac{I_{02}V_2 \, \text{Sinc}\left[\dfrac{(M\omega_m-\omega_{b2})T_m}{2}\right]+I_{03}V_3 \, \text{Sinc}\left[\dfrac{(M\omega_m-\omega_{b3})T_m}{2}\right]}{I_{01}V_1 \, \text{Sinc}\left[\dfrac{(M\omega_m-\omega_{b1})T_m}{2}\right]}\right\},
\end{aligned}
$$

(9-44)

where P_{1M}, P_{2M}, and P_{3M} are the amplitudes of the phasors P_{1M}, P_{2M}, and P_{3M}, respectively, as shown in Figure 9.9(b). Considering $P_{1M} \gg P_{2M} \gg P_{3M}$, Equation (9-44) can be simplified as

$$
\begin{aligned}
(\delta\phi_{b01})_{extr} &\approx \frac{P_{2M}+P_{3M}}{P_{1M}} \\
&= \frac{I_{02}V_2 \, \text{Sinc}\left[\dfrac{(M\omega_m-\omega_{b2})T_m}{2}\right]+I_{03}V_3 \, \text{Sinc}\left[\dfrac{(M\omega_m-\omega_{b3})T_m}{2}\right]}{I_{01}V_1 \, \text{Sinc}\left[\dfrac{(M\omega_m-\omega_{b1})T_m}{2}\right]}.
\end{aligned}
$$

(9-45)

Similar to case of the double-sensor multiplexed fiber-optic FMCW displacement sensor, the maximum phase error $(\delta\phi_{b01})_{max}$ caused by cross talk can be written as

$$(\delta\phi_{b01})_{max} = \arcsin\left\{\frac{I_{02}V_2 \, \text{Sinc}\left[\dfrac{(M\omega_m - \Omega_{b2})T_m}{2}\right] + I_{03}V_3 \, \text{Sinc}\left[\dfrac{(M\omega_m - \Omega_{b3})T_m}{2}\right]}{I_{01}V_1 \, \text{Sinc}\left[\dfrac{(M\omega_m - \Omega_{b1})T_m}{2}\right]}\right\}$$

$$\approx \frac{I_{02}V_2 \, \text{Sinc}\left[\dfrac{(M\omega_m - \Omega_{b2})T_m}{2}\right] + I_{03}V_3 \, \text{Sinc}\left[\dfrac{(M\omega_m - \Omega_{b3})T_m}{2}\right]}{I_{01}V_1 \, \text{Sinc}\left[\dfrac{(M\omega_m - \Omega_{b1})T_m}{2}\right]}$$

(9-46)

where Ω_1, Ω_2, and Ω_3 are the maximum angular frequencies of the three beat signals. The maximum displacement error $(\delta\Delta d_1)_{max}$ can be written as

$$(\delta\Delta d_1)_{max} = \frac{\lambda_0}{4\pi n}\arcsin\left\{\frac{I_{02}V_2\text{Sinc}\left[\dfrac{(M\omega_m - \Omega_{b2})T_m}{2}\right] + I_{03}V_3\text{Sinc}\left[\dfrac{(M\omega_m - \Omega_{b3})T_m}{2}\right]}{I_{01}V_1\text{Sinc}\left[\dfrac{(M\omega_m - \Omega_{b1})T_m}{2}\right]}\right\}$$

$$\approx \frac{\lambda_0\left\{I_{02}V_2\text{Sinc}\left[\dfrac{(M\omega_m - \Omega_{b2})T_m}{2}\right] + I_{03}V_3\text{Sinc}\left[\dfrac{(M\omega_m - \Omega_{b3})T_m}{2}\right]\right\}}{4\pi n I_{01}V_1\text{Sinc}\left[\dfrac{(M\omega_m - \Omega_{b1})T_m}{2}\right]}$$

(9-47)

where λ_0 is the central optical wavelength in free space.

If the dynamic range of the displacement sensor equals the modulation frequency ω_m (i.e., $\Omega_{b1} = M\omega_m + \omega_m/2$, $\Omega_{b2} = N\omega_m + \omega_m/2$, $\Omega_{b3} = P\omega_m + \omega_m/2$, where M, N, and P are integer), the maximum phase error due to the signal cross talk would be

$$(\delta\phi_{b01})_{max} = \frac{I_{02}V_2}{I_{01}V_1[2(N-M)+1]} + \frac{I_{03}V_3}{I_{01}V_1[2(P-M)+1]},$$

(9-48)

and the corresponding maximum displacement error would be

$$(\delta\Delta d_1)_{max} = \frac{\lambda_0}{4\pi n}\left[\frac{I_{02}V_2}{I_{01}V_1[2(N-M)+1]} + \frac{I_{03}V_3}{I_{01}V_1[2(P-M)+1]}\right].$$

(9-49)

Table 9.2 The maximum phase and displacement errors in the triple-sensor multiplexed single-mode fiber FMCW displacement sensor. (Assuming $\Omega_{b1} = M\omega_m+\omega_m/2$, $\Omega_{b2} = N\omega_m+\omega_m/2$, $\Omega_{b3} = P\omega_m+\omega_m/2$, $I_{01}V_1 = I_{02}V_2 = I_{03}V_3$, $n = 1$.)

$N-M$, $P-N$ (ω_m)	1	2	3	4	5
$(\delta\phi_{b01})_{max}$ (radian)	8/15	14/45	20/91	26/153	32/231
$(\delta\Delta l_1)_{max}$ (λ_0)	$2/15\pi$	$7/90\pi$	$5/91\pi$	$13/306\pi$	$8/231\pi$

Table 9.2 shows the maximum phase errors and the maximum displacement errors corresponding to different frequency gaps.

The maximum number of sensors that a frequency-division multiplexed single-mode fiber FMCW displacement sensor can have depends on both the cross talk and the contrast of the individual beat signals. The latter is determined by the laser-power budget, optical path difference, and coherence length of the laser.

A further increase in the number of sensors in the frequency-division multiplexed FMCW displacement sensor, however, is not practical because of the problems of high cross talk, poor contrast, and lengthy leading fiber. A feasible way to increase the number of sensors is a combination of frequency-division multiplexing and intensity-division multiplexing.

Figure 9.10 shows a four-sensor multiplexed single-mode fiber FMCW displacement sensor that consists of two independent double-sensor single-mode fiber FMCW displacement sensors [54]. The advantage of this scheme is

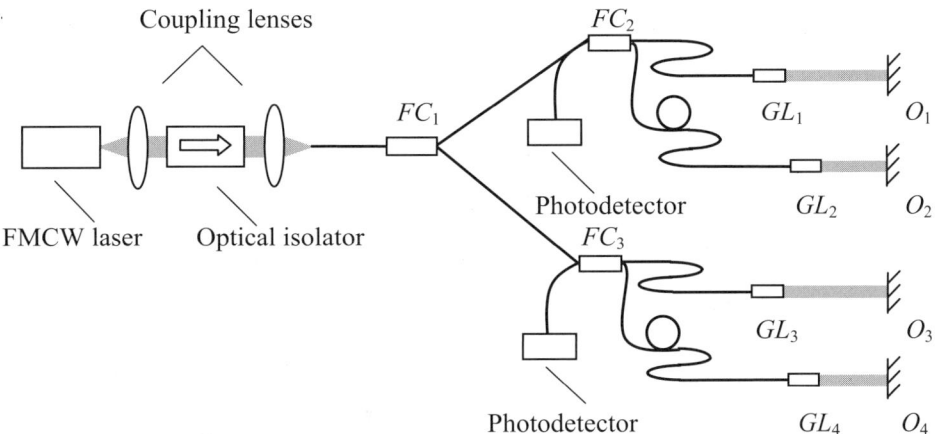

Figure 9.10. Four-sensor multiplexed single-mode fiber FMCW displacement sensor. (After [54].)

that it provides more sensors with less cross talk and avoids using lengthy leading fibers.

9.3 Fiber-optic FMCW Interferometric Strain Sensors

Strain and stress measurements are established, but active, topics in mechanical engineering. Many fields of industry, such as aerospace, aviation, and civil construction, employ various strain and stress sensors to ensure material and structural safety. However, the conventional electrical strain and stress sensors, such as metal-film strain sensors, are hardly likely to meet all the practical requirements. During the 1980s, researchers and engineers began to use fiber-optic sensors to measure the strain and stress in materials and structures, and some progress has been made. Now, the fast-developing fiber-optic strain and stress sensor technology has led to a new branch of engineering—smart materials and structures. Fiber-optic strain and stress sensors, like "fiber-optic nerves", can be either embedded in materials or surface-mounted on component parts. With this technology, future structures will become safer and more reliable under various load conditions.

Fiber-optic strain and stress sensors have been shown to possess a number of advantages over conventional electrical strain and stress sensors, such as immunity from electromagnetic interference, electrical isolation, chemical passivity, higher sensitivity, light weight, versatility, and long gauge length. Moreover, fiber-optic strain and stress sensors can be modified or combined to create fiber-optic strain and stress sensor networks—the multiplexed fiber-optic strain and stress sensors or the distributed fiber-optic strain and stress sensors.

So far, many types of fiber-optic strain and stress sensors have been proposed, so we will focus on fiber-optic FMCW interferometric strain and stress sensors. Specifically, we will discuss birefringent fiber FMCW interferometric strain sensors in this section and discuss birefringent fiber FMCW interferometric stress sensors in the next section.

9.3.1 Behavior of Birefringent Fibers under Tensile Forces

We know that the birefringent fiber supports two orthogonal polarization modes with different propagation numbers. This implies that a birefringent fiber itself could be a double-beam interferometer, in which one mode propagates the reference beam and another mode propagates the signal beam. One of the important advantages of this arrangement is that the reference arm of the fiber-optic interferometer, which usually makes the sensor configuration complex and produces an extra measurement error due to the temperature or strain on it, can be eliminated.

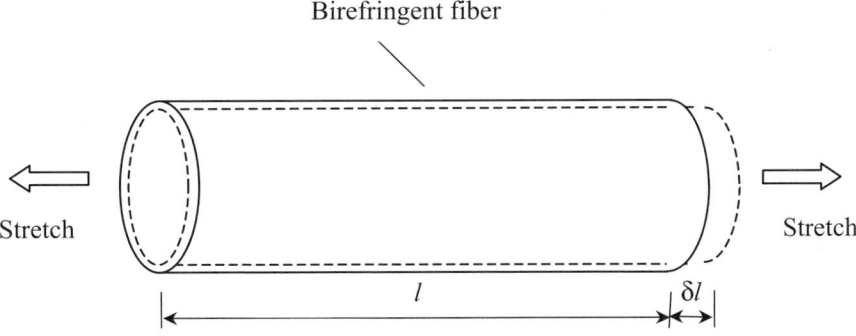

Figure 9.11. Distortion of an optical fiber under a tensile force.

When a birefringent fiber is stretched, the length of the fiber will increase (strain effect), the diameter of the fiber will get smaller (Poisson effect), and the optical refractive index of the fiber will change slightly (elasto-optic effect), as shown in Figure 9.11. Theoretically, stretching a birefringent fiber does not destroy the birefringence in the fiber because the optical axis of any such stretch-induced birefringence is perpendicular to the principal axes of the birefringent fiber. However, stretching a birefringent fiber may change the optical path difference between the two orthogonal polarization modes.

For instance, the original optical path difference *OPD* between the two orthogonal polarization modes in a birefringent fiber can be written as

$$OPD = (n_{ex} - n_{ey})l \tag{9-50}$$

where n_{ex} is the effective refractive index of the $\mathrm{HE_{11}}^x$ mode, n_{ey} is the effective refractive index of the $\mathrm{HE_{11}}^y$ mode, and l is the length of the birefringent fiber. When the fiber is stretched, the variation of the optical path difference $\delta(OPD)$ will be

$$\delta(OPD) = (n_{ex} - n_{ey})\delta l + \delta(n_{ex} - n_{ey})l \tag{9-51}$$

where δl is the extension of the fiber, $\delta(n_{ex} - n_{ey})$ is the variation of the effective refractive-index difference due to the fiber diameter reduction and the refractive-index variation. In general, $\delta(n_{ex} - n_{ey})$ is relatively small, so Equation (9-51) can be simplified as

$$\delta(OPD) = (n_{ex} - n_{ey})\delta l \tag{9-52}$$

The polarization state of light generally is described by using the Jones vector (a vector containing two orthogonal polarization components), which usually is graphically represented by the Lissajous pattern, or by using the Stokes vector (a four-element vector), which usually is represented graphically by the Poincarè sphere. Figure 9.12 illustrates the real variation of the polarization state of the output beam from a continuously stretched elliptical-core birefringent fiber whenever a linearly polarized laser beam is coupled into only one polarization mode. The Stokes vector of the output beam constantly locates nearly the same point on the equator of the Poincarè sphere. This indicates that the polarization state of the output beam does not change but remains linear. In other words, stretching a birefringent fiber does not produce any mode coupling.

Figure 9.13 demonstrates the variation of the polarization state of the output beam from a continuously stretched elliptical-core birefringent fiber whenever the input beam is coupled into two polarization modes with approximately equal intensities. The polarization state of the output beam in this case periodically changes from a linear state (Figure 9.13(a)) to an elliptical state(b), a circular state (c), an elliptical state (d), to another linear state (e) that is perpendicular to the first one, and so forth.

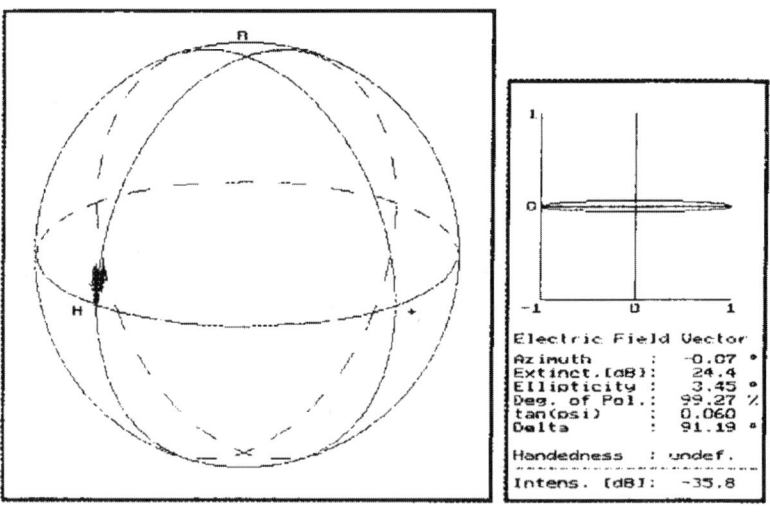

(a) The Poincarè sphere representation.

(b) The Lissajous pattern representation.

Figure 9.12. Variation of the polarization state of the output beam from a continuously stretched elliptical-core birefringent fiber when the input beam is launched into one polarization mode.

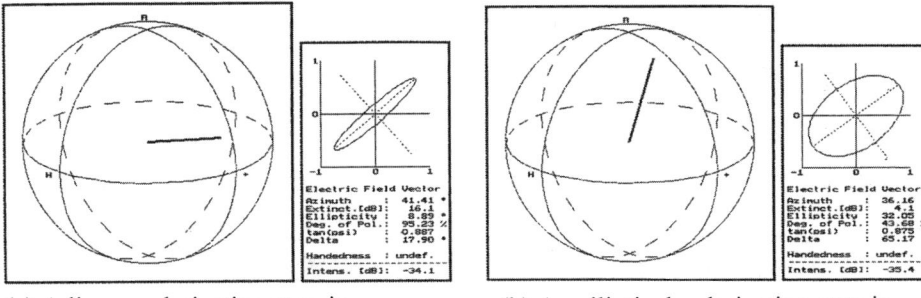

(a) A linear polarization state in quadrants I-III.

(b) An elliptical polarization state in quadrants I-III.

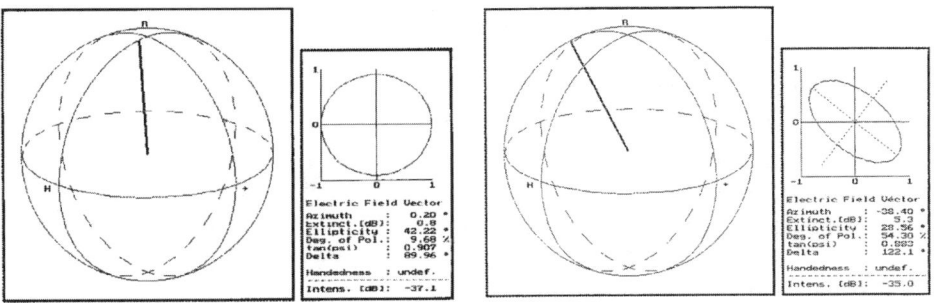

(c) A circular polarization state.

(d) An elliptical polarization state in quadrants II-IV.

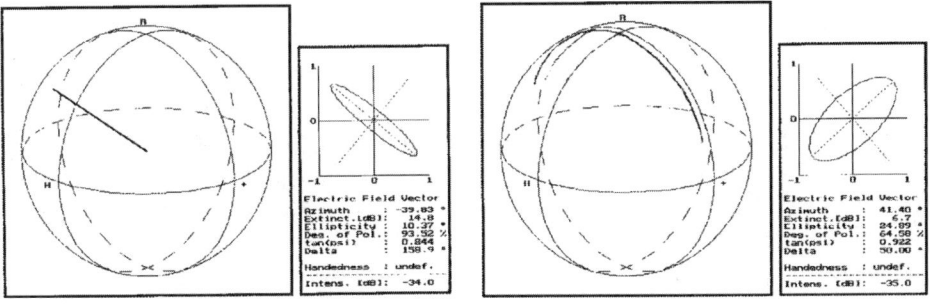

(e) A linear polarization state in quadrants II-IV.

(f) The trace of the Stokes vector on the Poincarè sphere.

Figure 9.13. Variation of the polarization state of the output beam from a continuously stretched elliptical-core birefringent fiber when the input beam is launched into two polarization modes.

Obviously, stretching a birefringent fiber does change the optical path difference between the two orthogonal polarization modes. The experimental data show that the fiber extension period, which corresponds to a full period of the polarization state variation or a full wavelength variation of optical path difference, is 1.50 mm whenever the experiment is performed with a 3 μm×1.5 μm elliptical-core fiber and a 0.660 μm laser diode. Hence, birefringent fibers can be used for constructing fiber-optic strain sensors.

9.3.2 Transmissive Birefringent Fiber FMCW Strain Sensors

Figure 9.14(a) shows schematically a transmissive intrinsic birefringent fiber FMCW interferometric strain sensor. The sensor consists of two polarizers (P_1 and P_2) and three lengths of birefringent fiber linked by two fiber-optic connectors (FN_1 and FN_2). Two lengths of birefringent fiber are used for laser power transmission (called lead-in fiber and lead-out fiber, respectively), while the third length of birefringent fiber, located in the middle, is used as a strain-sensing fiber. The principal axes of the sensing fiber are rotated 45° with respect to the principal axes of both the lead-in and lead-out fibers.

A polarized FMCW laser beam is coupled into the lead-in fiber in one of the orthogonal polarization modes (say the $HE_{11}{}^x$ mode). When the beam enters the sensing fiber, it will separate into two beams because of the 45° rotation of the principal axes. Similarly, these beams will separate again when they enter the lead-out fiber. Therefore, there will be four beams emerging from the lead-out fiber, which can be named the $HE_{11}{}^x$-$HE_{11}{}^x$-$HE_{11}{}^x$ mode beam, the $HE_{11}{}^x$-$HE_{11}{}^x$-$HE_{11}{}^y$ mode beam, the $HE_{11}{}^x$-$HE_{11}{}^y$-$HE_{11}{}^x$ mode beam, and the $HE_{11}{}^x$-$HE_{11}{}^y$-$HE_{11}{}^y$ mode beam, according to the mode difference in different fibers, as shown in Figure 9.14(b).

The second polarizer P_2 located in the front of a photodetector is aligned in parallel with the principal axis of one polarization mode (say the $HE_{11}{}^y$ mode), so that only two beams, the $HE_{11}{}^x$-$HE_{11}{}^x$-$HE_{11}{}^y$ mode beam and the $HE_{11}{}^x$-$HE_{11}{}^y$-$HE_{11}{}^y$ mode beam, can pass through P_2 and reach a photodetector.

Obviously, the two interfering beams travel only in the sensing fiber with different propagation numbers. Therefore, the optical path difference between the two interfering beams depends only on the parameters of the sensing fiber, not on the parameters of the lead-in and lead-out fibers. This means that the lead-in and lead-out fibers are insensitive to the environment even if they are warmed or stretched.

The optical path difference OPD between the $HE_{11}{}^x$-$HE_{11}{}^x$-$HE_{11}{}^y$ mode beam and the $HE_{11}{}^x$-$HE_{11}{}^y$- $HE_{11}{}^y$ mode beam can be written as

$$OPD = (n_{ex} - n_{ey})l ,$$

(9-53)

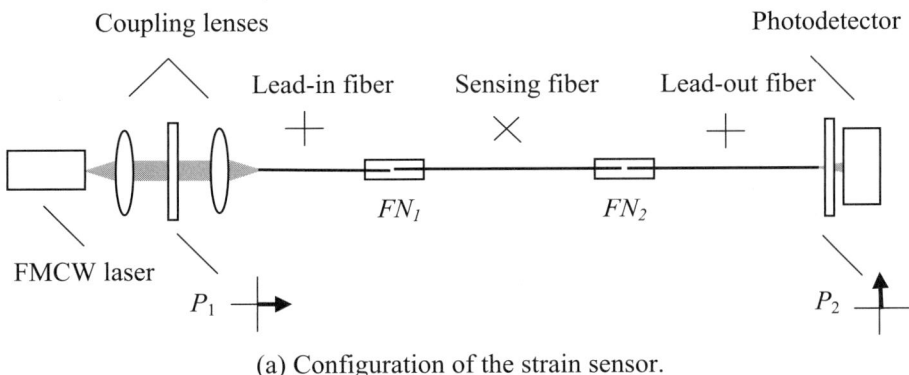

(a) Configuration of the strain sensor.

	HE_{11}^{x}-HE_{11}^{y}	HE_{11}^{x}-HE_{11}^{y}- HE_{11}^{y}
		HE_{11}^{x}- HE_{11}^{x}- HE_{11}^{y}
		HE_{11}^{x}-HE_{11}^{y}-HE_{11}^{x} (Blocked)
HE_{11}^{x}	HE_{11}^{x}-HE_{11}^{x}	HE_{11}^{x}-HE_{11}^{x}-HE_{11}^{x} (Blocked)

| Lead-in fiber | Strain sensing fiber | Lead-out fiber |

(b) Propagation of the laser beams in the birefringent fibers.

Figure 9.14. Transmissive birefringent fiber FMCW strain sensor.

where n_{ex} is the effective refractive index of the HE_{11}^{x} mode, n_{ey} is the effective refractive index of the HE_{11}^{y} mode, and l is the length of the sensing fiber.

In the case of sawtooth-wave modulation, for instance, the intensity of the beat signal in a modulation period $I(t)$ can be written as

$$I(OPD,t) = I_0\left[1 + V\cos\left(\frac{2\pi\Delta\nu\nu_m OPD}{c}t + \frac{2\pi}{\lambda_0}OPD\right)\right]$$

$$= I_0[1 + V\cos(2\pi\nu_b t + \phi_{b0})] \qquad , \qquad (2\text{-}63)$$

where I_0 is the average intensity of the beat signal, V is the contrast of the beat signal, $\Delta\nu$ is the optical frequency modulation excursion, ν_m is the modulation frequency, c is the speed of light in free space, λ_0 is the central optical wavelength in free space, and ν_b and ϕ_{b0} are the frequency and initial phase of the beat signal, respectively. Considering Equation (9-53), the frequency of the beat signal v_b can be written as

$$v_b = \frac{(n_{ex} - n_{ey})\Delta v v_m l}{c},$$

(9-54)

and the initial phase of the beat signal ϕ_{b0} can be written as

$$\phi_{b0} = \frac{2\pi(n_{ex} - n_{ey})l}{\lambda_0}.$$

(9-55)

If the sensing fiber is stretched, the variations of the frequency and initial phase of the beat signal, Δv_b and $\Delta \phi_{b0}$, will be

$$\Delta v_b = \frac{(n_{ex} - n_{ey})\Delta v v_m \Delta l}{c},$$

(9-56)

$$\Delta \phi_{b0} = \frac{2\pi(n_{ex} - n_{ey})\Delta l}{\lambda_0},$$

(9-57)

where Δl is the elongation of the sensing fiber. Hence, the strain of the sensing fiber ε ($\varepsilon = \Delta l/l$) can be calculated by using either

$$\varepsilon = \frac{c}{(n_{ex} - n_{ey})\Delta v v_m l}\Delta v_b,$$

(9-58)

or

$$\varepsilon = \frac{\lambda_0}{2\pi(n_{ex} - n_{ey})l}\Delta \phi_{b0}.$$

(9-59)

Figure 9.15 shows another version of the transmissive intrinsic birefringent fiber FMCW interferometric strain sensor, which comprises a single long length of birefringent fiber [71]. Two polarization mode couplers (MC_1 and MC_2) located in the middle separate the whole fiber into three sections. The first one, from the launching end of the fiber to MC_1, is used as a lead-in fiber; the second one, between MC_1 and MC_2, is used as a strain-sensing fiber; and the last one, from MC_2 to the emerging end of the fiber, is used as a lead-out fiber.

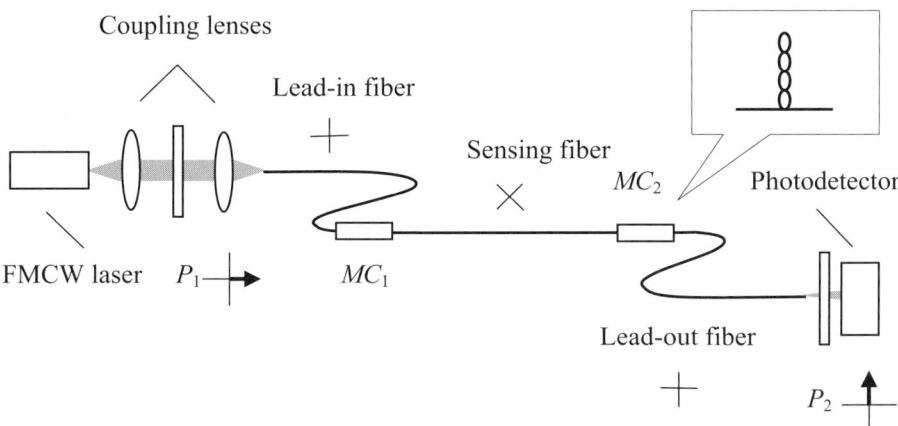

Figure 9.15. Transmissive single birefringent fiber FMCW strain sensor. (After [71].)

The polarization mode coupler, which can be a spring clip pressing on the birefringent fiber or simply realized by twisting and bending a short length of the birefringent fiber (typically about 5 cm), couples some laser energy from one polarization mode into another. Therefore, as in the previous configuration, if a polarized FMCW laser beam is coupled into the birefringent fiber in one polarization mode, there will be four beams emerging from the lead-out fiber. The second polarizer P_2, located in the front of a photodetector, is aligned parallel to the principal axis of one of the polarization modes so that only two beams, which propagate through the sensing fiber with different propagation numbers, can pass through P_2 and produce a beat signal.

The advantage of the second version is that the lengths of the lead-in fiber, sensing fiber, and lead-out fiber are changeable simply by releasing the polarization mode couplers and then reloading them at different locations on the birefringent fiber.

9.3.3 Reflectometric Birefringent Fiber FMCW Strain Sensors

Figure 9.16(a) shows schematically a reflectometric intrinsic birefringent fiber FMCW interferometric strain sensor, which consists of two lengths of birefringent fiber linked by a fiber-optic connector (*FN*) and a polarizer (*P*) located in front of a photodetector [80]. The first length of birefringent fiber is used as a leading fiber, while the second length of birefringent fiber is used as a sensing fiber whose principal axes are rotated at 45° with respect to the leading fiber and whose distant end is attached with a small mirror (*M*). (Usually the cleaved end of the fiber is coated with silver and thus acts as a mirror.)

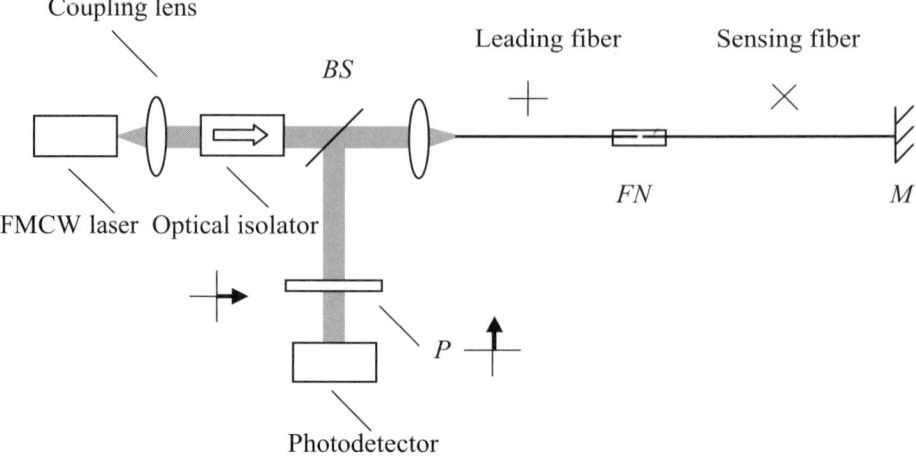

(a) Configuration of the strain sensor.

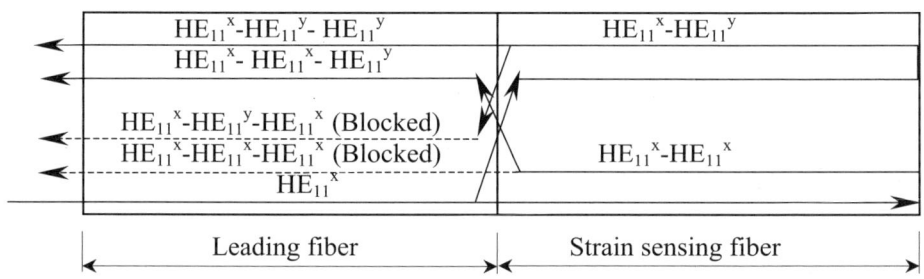

(b) Propagation of the laser beams in the birefringent fibers.

Figure 9.16. Reflective birefringent fiber FMCW strain sensor. (After [80].)

A polarized FMCW laser beam is coupled into the leading fiber in one po-larization mode (say HE_{11}^x mode). After entering the sensing fiber, the incident beam separates into two beams, the HE_{11}^x mode beam and the HE_{11}^y mode beam, because the principal axis of the sensing fiber is rotated 45° with respect to the leading fiber. These two beams propagate along the sensing fiber, reflect back on M, and separate into two again when they pass through FN in the op-posite direction. Therefore, there will be four beams emerging from the fiber, which can be named the HE_{11}^x-HE_{11}^x-HE_{11}^x mode beam, the HE_{11}^x-HE_{11}^x-HE_{11}^y mode beam, the HE_{11}^x-HE_{11}^y-HE_{11}^x mode beam, and the HE_{11}^x-HE_{11}^y-HE_{11}^y mode beam, according to the mode difference in different fibers at different times, as shown in Figure 9.16(b). These four beams are reflected by BS and propagate toward P. Since the transmission direction of P is aligned parallel to the principal axis of one polarization mode (say $HE_{11}{}^y$), only two beams, the $HE_{11}{}^x$-$HE_{11}{}^x$-$HE_{11}{}^y$ mode beam and the $HE_{11}{}^x$-$HE_{11}{}^y$-$HE_{11}{}^y$ mode beam, can pass through P and coherently interfere to produce a beat signal, which is finally received by a photodetector.

The optical path difference OPD between the two interfering beams can be written as

$$OPD = 2(n_{ex} - n_{ey})l$$,
(9-60)

where n_{ex} is the effective refractive index of the $HE_{11}{}^x$ mode, n_{ey} is the effective refractive index of the $HE_{11}{}^y$ mode, and l is the length of the sensing fiber.

Similarly, the optical path difference between the two interfering beams is proportional to the length of the sensing fiber but not related to the leading fiber. Moreover, since the optical path difference in the reflectometric birefringent fiber strain sensor is twice that of the transmissive one, the resolution of the reflectometric strain sensor is doubly improved.

In the case of sawtooth-wave modulation, for instance, the intensity of the beat signal in a modulation period $I(t)$ can be written as

$$I(OPD,t) = I_0\left[1 + V\cos\left(\frac{2\pi\Delta\nu\nu_m OPD}{c}t + \frac{2\pi}{\lambda_0}OPD\right)\right]$$
$$= I_0[1 + V\cos(2\pi\nu_b t + \phi_{b0})]$$,
(2-63)

where I_0 is the average intensity of the beat signal, V is the contrast of the beat signal, $\Delta\nu$ is the optical frequency modulation excursion, ν_m is the modulation frequency, c is the speed of light in free space, λ_0 is the central optical wavelength in free space, and ν_b and ϕ_{b0} are the frequency and initial phase of the beat signal, respectively. Considering Equation (9-60), the frequency of the beat signal ν_b can be written as

$$\nu_b = \frac{2(n_{ex} - n_{ey})\Delta\nu\nu_m l}{c}$$,
(9-61)

and the initial phase of the beat signal ϕ_{b0} can be written as

$$\phi_{b0} = \frac{4\pi(n_{ex} - n_{ey})l}{\lambda_0}$$.
(9-62)

If the sensing fiber is stretched, the variations of the frequency and initial phase of the beat signal Δv_b and $\Delta \phi_{b0}$ will be

$$\Delta v_b = \frac{2(n_{ex} - n_{ey})\Delta v v_m \Delta l}{c} ,$$

(9-63)

$$\Delta \phi_{b0} = \frac{4\pi(n_{ex} - n_{ey})\Delta l}{\lambda_0} ,$$

(9-64)

where Δl is the elongation of the sensing fiber. Hence, the strain of the sensing fiber ε can be given by

$$\varepsilon = \frac{c}{2(n_{ex} - n_{ey})\Delta v v_m l} \Delta v_b ,$$

(9-65)

or

$$\varepsilon = \frac{\lambda_0}{4\pi(n_{ex} - n_{ey})l} \Delta \phi_{b0} .$$

(9-66)

Figure 9.17 shows another version of the reflectometric intrinsic birefringent fiber FMCW interferometric strain sensor, which comprises a single length of birefringent fiber attached with a mirror [77]. A polarization mode coupler (MC) located in the middle of the birefringent fiber divides it into two sections. The first section, extending from the launch end to MC, is used as leading fiber, while the second section, extending from MC to the mirrored end, is used as a strain-sensing fiber.

Similarly, a polarized FMCW laser beam is coupled into the leading fiber in one polarization mode (say the HE_{11}^x mode) and divided into two beams by MC. These two beams are reflected back by the mirror at the end of the fiber and divided again by MC. Therefore, there are four beams exiting from the input end of the birefringent fiber. Choosing any two beams with the same polarization state by using a polarizer P, we can determine the strain on the sensing fiber.

The advantage of the second version of the reflectometric intrinsic birefringent fiber FMCW strain sensor is that the lengths of the leading fiber and the sensing fiber are changeable. In other words, we can simply release the polarization mode couplers and then reload them at different locations to change lengths of the leading fiber and the sensing fiber. (Of course, the total length of the fibers is the same.)

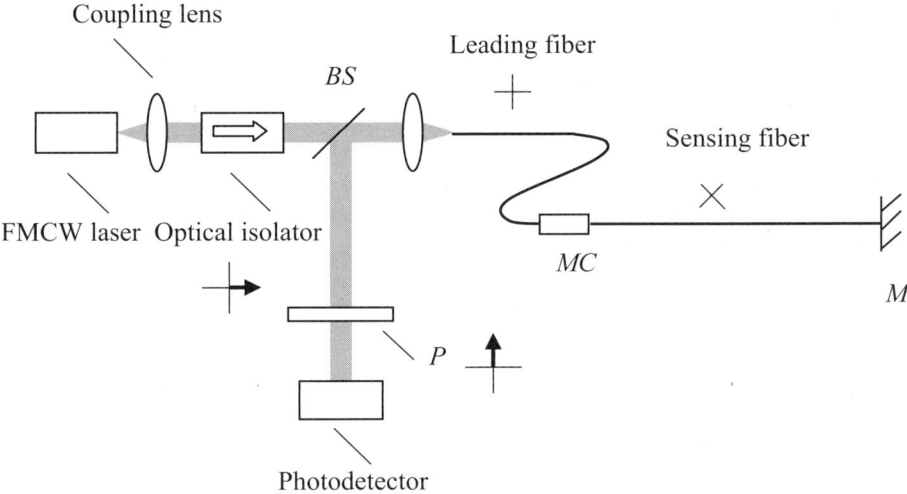

Figure 9.17. Reflectometric single birefringent fiber FMCW strain sensor. (After [77].)

The birefringent fiber FMCW strain sensors discussed previously (including transmissive configurations and reflectometric configurations) can have a long environment-insensitive leading fiber and a long strain-sensing fiber. The former is due to the fact that the two interfering beams propagate along the leading fiber in the same mode. The latter is because of the fact that the optical path difference of the two interfering beams in the birefringent sensing fiber is proportional to the difference of the effective refractive indexes ($n_{ex}-n_{ey}$) of the two orthogonal polarization modes rather than the effective refractive index n_e of a signal-mode fiber.

For instance, for the reflectometric birefringent fiber FMCW interferometric strain sensors, assuming that the coherence length of an FMCW laser is 10 meters and that the difference of the effective refractive indexes ($n_{ex}-n_{ey}$) of the birefringent fiber is 0.0005, the maximum length of the birefringent sensing fiber could be up to 10 kilometers (if the attenuation is acceptable). By comparison, the maximum length of the single-mode sensing fiber (see Figure 7.23) can be only about 3.3 meters, assuming the effective refractive index n_e of the signal-mode fiber is 1.5. Hence, birefringent fiber FMCW strain sensors are well-suited for remotely measuring the strain on huge constructions, such as oil tanks, suspension bridges, or dams.

9.4 Fiber-optic FMCW Interferometric Stress Sensors

9.4.1 Behaviour of Birefringent Fibers under Perpendicular Forces

In an elliptical-core birefringent fiber with a step-index profile, for instance, the electric field in each mode can be described with a series of Mathieu functions [10]. On the assumption of little eccentricity and operation at the single-mode threshold (i.e., the normalized frequency V = 2.4), the propagation-number difference of the two orthogonal polarization modes $\Delta\beta$ can be written as

$$\Delta\beta \approx \frac{e^2}{a}\left(\frac{\Delta}{2}\right)^{3/2} ,$$

(9-67)

where e is the eccentricity of the fiber core ($e^2 = [1-(b/a)^2]$, where a is the semimajor axis and b is the semiminor axis) and Δ is the relative refractive-index difference, as shown in Figure 9.18.

When a birefringent fiber is pressed, a new linear birefringence will be introduced (the elasto-optic effect). Assuming that the pressure-induced birefringence is much larger than the asymmetry-induced birefringence, the resulting birefringence will still be linear, but the optical axis of the pressed fiber will be almost parallel to the direction of the pressure, as shown in Figure 9.19. In other words, the principal axes of the pressed fiber will rotate at an angle that depends on the direction of the pressure.

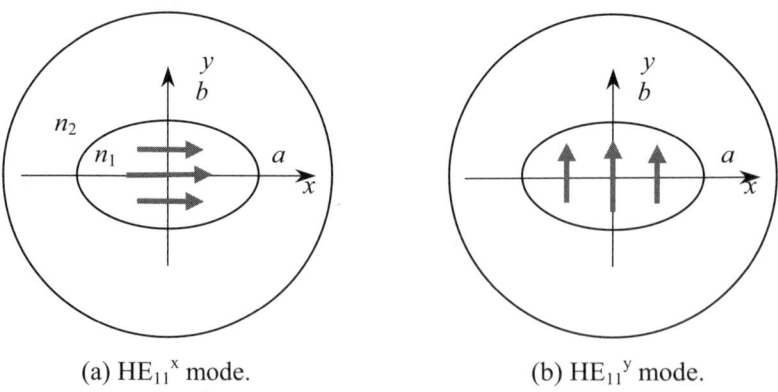

(a) HE_{11}^x mode.   (b) HE_{11}^y mode.

Figure 9.18. Polarization modes in elliptical-core birefringent fiber.

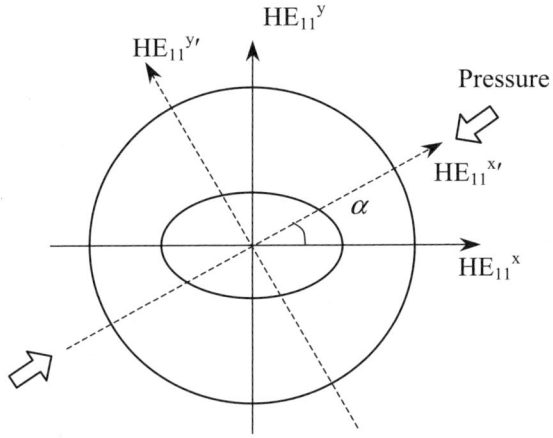

Figure 9.19. Birefringence variation in a stressed elliptical-core fiber.

If a polarized laser beam is originally launched into one of the orthogonal polarization modes (say $HE_{11}{}^x$ mode) and the laser beam enters the pressed fiber section, it will separate into two new components, the $HE_{11}{}^{x'}$ mode beam and the $HE_{11}{}^{y'}$ mode beam, and these beams will propagate through the pressed fiber with the new, different propagation numbers along the new principal axes. When the two beams return to the ordinary birefringent fiber, they will reform and propagate in the original $HE_{11}{}^x$ and $HE_{11}{}^y$ modes. Apparently, the pressed fiber, acting as a polarization mode coupler, can couple some light energy from the $HE_{11}{}^x$ mode into the $HE_{11}{}^y$ mode.

The pressed fiber can be treated as a wave plate, because it has its own principal axes and effective refractive indexes. The intensity of the light beam I_y that is coupled from the $HE_{11}{}^x$ mode to the $HE_{11}{}^y$ mode can be written as

$$I_y = I_x \sin^2 2\alpha \sin^2 \frac{\phi}{2},$$

(9-68)

where I_x is the intensity of the incident $HE_{11}{}^x$ mode beam, α is the angle between $HE_{11}{}^{x'}$ and $HE_{11}{}^x$, and ϕ is the optical phase difference between the $HE_{11}{}^{x'}$ mode beam and $HE_{11}{}^{y'}$ mode beam, which is given by

$$\phi = \frac{2\pi}{\lambda_0}(n_{ex}' - n_{ey}')l$$

$$= \frac{2\pi}{\lambda_0} C_B \cdot \frac{F}{a},$$

(9-69)

where λ_0 is the wavelength in free space, n_{ex}' and n_{ey}' are effective refractive indexes of the new orthogonal modes in the pressed fiber, C_B is the Brewster constant, F is the vertical force, and a is the effective width of the fiber.

Pressing a birefringent fiber can couple some light energy from one polarization mode into another. Therefore, this fact can be employed (i) to make polarization mode couplers and (ii) to construct distributed birefringent fiber stress sensors.

9.4.2 Transmissive Distributed Birefringent Fiber FMCW Stress Sensors

The schematic diagram shown in Figure 9.20(a) represents a transmissive distributed intrinsic birefringent fiber FMCW interferometric stress sensor [23]. An FMCW laser beam is polarized by the first polarizer (P_1), and then is coupled into a length of birefringent fiber in one polarization mode (say the HE_{11}^x mode). The stress to be measured causes coupling of part of the incident beam into another orthogonal polarization mode (i.e., the HE_{11}^y mode) due to the elasto-optic effect. Therefore, there will be two beams, the coupled HE_{11}^y mode beam and the left HE_{11}^x mode beam, propagating along the fiber after coupling. The second polarizer (P_2), whose transmission direction is aligned at an angle θ with respect to the HE_{11}^x mode, is placed in front of a photodetector to collect some energy from both the HE_{11}^x mode and HE_{11}^y mode beams to interfere.

From Figure 9.20(b), it can be seen that the optical path difference OPD between the two interfering beams equals

$$OPD = (n_{ex} - n_{ey})z , \qquad (9\text{-}70)$$

where n_{ex} is the effective refractive index of the HE_{11}^x mode, n_{ey} is the effective refractive index of the HE_{11}^y mode, and z is the fiber length from the stress position to the distant end of the fiber.

In the case of sawtooth-wave modulation, for instance, the intensity of the beat signal in a modulation period $I(t)$ can be written as

$$I(OPD, t) = I_0 \left[1 + V \cos \left(\frac{2\pi \Delta v v_m OPD}{c} t + \frac{2\pi}{\lambda_0} OPD \right) \right]$$
$$= I_0 [1 + V \cos(2\pi v_b t + \phi_{b0})] , \qquad (2\text{-}63)$$

where I_0 is the average intensity of the beat signal, V is the contrast of the beat signal, Δv is the optical frequency modulation excursion, v_m is the modulation frequency, c is the speed of light in free space, λ_0 is the central optical wave-

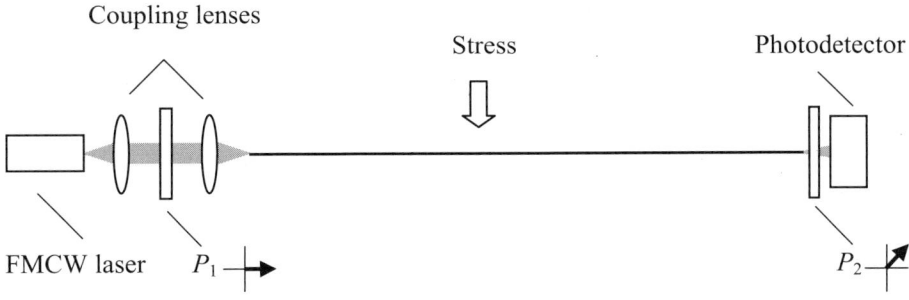

(a) Configuration of the strain sensor.

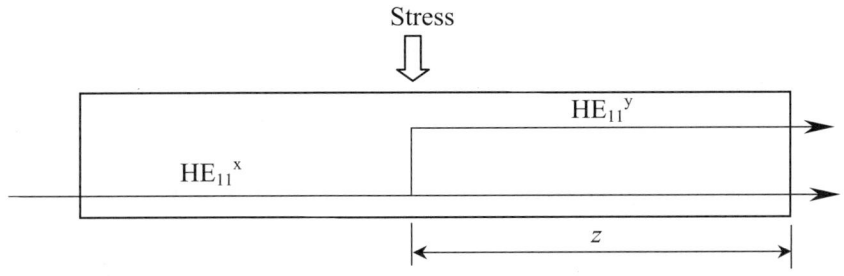

(b) Propagation of the laser beams in the birefringent fiber.

Figure 9.20. Transmissive distributed birefringent fiber FMCW stress sensor. (After [23].)

length in free space, and v_b and ϕ_{b0} are the frequency and initial phase of the beat signal, respectively. Considering Equation (9-70), the frequency of the beat signal v_b can be written as

$$v_b = \frac{(n_{ex} - n_{ey})\Delta v v_m z}{c}.$$

(9-71)

Therefore, the location of the stress can be determined by

$$z = \frac{c}{(n_{ex} - n_{ey})\Delta v v_m} \Delta v_b.$$

(9-72)

The intensity I_1 of the signal beam, which is split from the coupled $HE_{11}{}^y$ mode beam, can be written as

$$I_1 = I_i \zeta \sin^2 \theta,$$

(9-73)

where I_i is the intensity of the incident beam, ζ is the mode coupling coefficient, and θ is the angle between the transmission directions of the second polarizer and $HE_{11}{}^x$ mode, neglecting the attenuation of the laser beam in the fiber (the common factor for all beams). The intensity of the reference beam I_2, which is split from the left $HE_{11}{}^x$ mode beam, can be written as

$$I_2 = I_i(1-\zeta-\eta)\cos^2\theta,$$

(9-74)

where η is the coupling-loss coefficient.

The advantage of the transmissive distributed birefringent fiber FMCW stress sensor is that all the beams in the fiber are transmitted forward so that the influence of the feedback light on the laser is largely reduced. The drawback may be that the two interfering beams are normally not of equal intensity; thus, the signal contrast cannot be optimized.

9.4.3 Reflectometric Distributed Birefringent Fiber FMCW Stress Sensors

Figure 9.21(a) shows a reflectometric distributed intrinsic birefringent fiber FMCW interferometric stress sensor [73]. A polarized FMCW laser beam propagates through a beam splitter (*BS*) and then enters a long length of birefringent fiber in one polarization mode (say the $HE_{11}{}^x$ mode). At the distant end of the fiber, a tiny mirror (*M*) attaches to the fiber and reflects the incident laser beam back to the fiber. (Usually the cleaved end of the fiber is coated with silver and thus acts as a mirror.) A polarizer (*P*), whose transmission direction is aligned parallel to the principal axis of one polarization mode (say the $HE_{11}{}^y$ mode), is placed in front of a photodetector.

If a stress is applied to the fiber in any position along its length, energy coupling between the $HE_{11}{}^x$ mode and the $HE_{11}{}^y$ mode for both the incident beam and the reflected beam will occur. From Figure 9.21(b), it can be seen that there exist four beams that propagate back to the sensor system: B_1 is an reflected $HE_{11}{}^x$ mode beam, B_2 is an $HE_{11}{}^y$ mode beam that is coupled from the incident $HE_{11}{}^x$ mode beam, B_3 is an $HE_{11}{}^y$ mode beam that is coupled from the reflected $HE_{11}{}^x$ mode beam, and B_4 is an $HE_{11}{}^x$ mode beam that is coupled from the reflected coupled $HE_{11}{}^y$ mode beam. These four beams are reflected by the beam splitter *BS* and propagate toward *P*. However, because the transmission direction of *P* is aligned perpendicular to the principal axis of the $HE_{11}{}^x$ mode, only the two $HE_{11}{}^y$ mode beams B_2 and B_3 can pass through *P* and finally reach the photodetector.

Because the two $HE_{11}{}^y$ mode beams B_2 and B_3 are excited at different times and have different experiences in the fiber, there is an optical path difference between them when they interfere. The optical path difference *OPD* between B_2 and B_3 can be written as

$$OPD = 2(n_{ex} - n_{ey})z,$$
(9-75)

where n_{ex} is the effective refractive index of the $HE_{11}{}^{x}$ mode, n_{ey} is the effective refractive index of the $HE_{11}{}^{y}$ mode, and z is the fiber length between the points where the stress acts and the mirror end of the fiber.

In the case of sawtooth-wave modulation, for instance, the intensity of the beat signal in a modulation period $I(t)$ can be written as

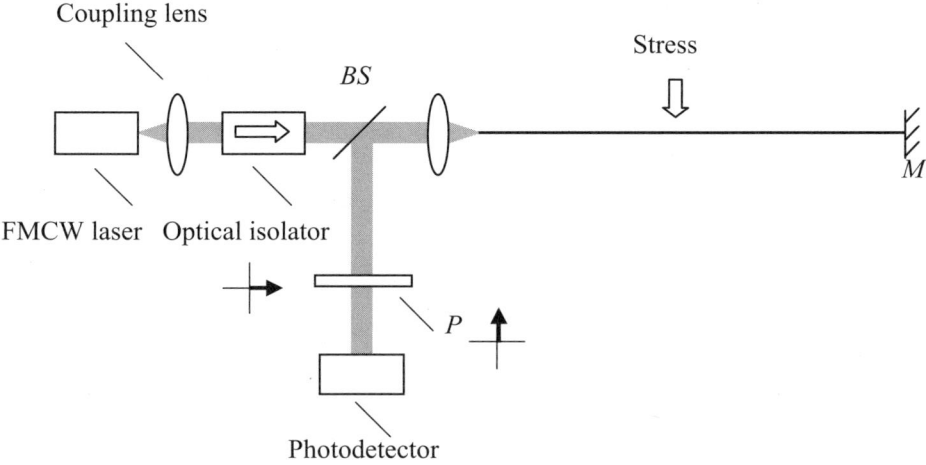

(a) Configuration of the stress sensor.

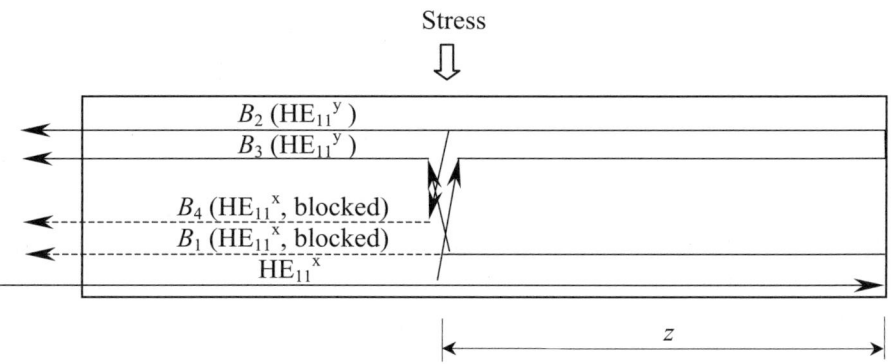

(b) Propagation of the laser beams in the birefringent fiber.

Figure 9.21. Reflectometric birefringent fiber FMCW distributed stress sensor. (After [73].)

$$I(OPD,t) = I_0 \left[1 + V \cos\left(\frac{2\pi\Delta\nu\,\nu_m OPD}{c}t + \frac{2\pi}{\lambda_0}OPD \right) \right]$$

$$= I_0[1 + V \cos(2\pi\nu_b t + \phi_{b0})]$$

$$(2\text{-}63)$$

where I_0 is the average intensity of the beat signal, V is the contrast of the beat signal, $\Delta\nu$ is the optical frequency modulation excursion, ν_m is the modulation frequency, c is the speed of light in free space, λ_0 is the central optical wavelength in free space, and ν_b and ϕ_{b0} are the frequency and initial phase of the beat signal, respectively. Considering Equation (9-75), the frequency of the beat signal ν_b can be written as

$$\nu_b = \frac{2(n_{ex} - n_{ey})\Delta\nu\,\nu_m z}{c} .$$

$$(9\text{-}76)$$

Therefore, the location of the stress can be determined by

$$z = \frac{c}{2(n_{ex} - n_{ey})\Delta\nu\,\nu_m}\nu_b .$$

$$(9\text{-}77)$$

The intensity I_2 of the beam B_2 can be written as

$$I_2 = I_i\zeta R(1 - \zeta - \eta) ,$$

$$(9\text{-}78)$$

where I_i is the intensity of the incident beam, ζ is the mode-coupling coefficient, η is the coupling-loss coefficient, and R is the reflectance of the mirror, neglecting the attenuation of the laser beam in the fiber (a common factor for all beams). The intensity I_3 of the beam B_3 can be written as

$$I_3 = I_i(1 - \zeta - \eta)R\zeta$$

$$= I_2 .$$

$$(9\text{-}79)$$

Obviously, the two HE_{11}^y mode beams always have equal intensities (the condition that ensures that the beat signal has a good contrast), and the intensity of the beat signal is directly proportional to the intensities of the coupled beams. In addition, the reflectometric nature of the sensor structure results in twice the measurement resolution, and the probe-type sensing fiber could be more adaptable to practical applications, such as intruder detection.

The advantage of the intrinsic birefringent fiber FMCW distributed stress sensors (including transmissive and reflectometric stress sensors) is that they can have a long length of sensing fiber (up to 10 kilometers). The limitation of the intrinsic birefringent fiber FMCW distributed stress sensors is that they are quasi-distributed rather than fully distributed, even though they can simultaneously measure a few stresses applied at different locations by using the frequency-division multiplexing method. Therefore, these sensors are suited to measuring single, or at best a few, stresses along the fiber at a time.

9.5 Fiber-optic FMCW Interferometric Temperature Sensors

Fiber-optic temperature sensors are particularly useful in electrical transformers, jet engines, and certain medical treatments because they are isolative in electricity, passive in chemistry, small in size, and light in weight. In the following subsections, we introduce a reflectometric single-mode fiber FMCW temperature sensor and a multiplexed reflectometric single-mode fiber FMCW temperature sensor.

9.5.1 Reflectometric Single-Mode Fiber FMCW Temperature Sensors

Figure 9.22 shows a reflectometric intrinsic single-mode fiber FMCW interferometric temperature sensor [39]. The sensor primarily consists of an X-type single-mode fiber coupler (*FC*) and a short length of single-mode fiber (typically about 5 mm) acting as the temperature-sensing probe. The temperature-sensing fiber is connected with one output fiber of the fiber coupler (usually called the leading fiber) through a narrow air gap to construct a single-mode fiber cavity.

An FMCW laser beam is separated by *FC* into two parts. One is detected by a photodetector (D_1) to monitor the laser power variation, and the other is partially reflected at the end of the leading fiber (Fresnel reflection), passes through the fiber cavity, and then undergoes a second reflection at the distant end. The two reflected beams mix coherently to produce a beat signal, which propagates back through the fiber coupler and is detected by another photodetector (D_2).

The optical path difference *OPD* between the two interfering beams is

$$OPD = 2n_e l,$$
(9-80)

where n_e is the effective refractive index of the single-mode fiber and l is the

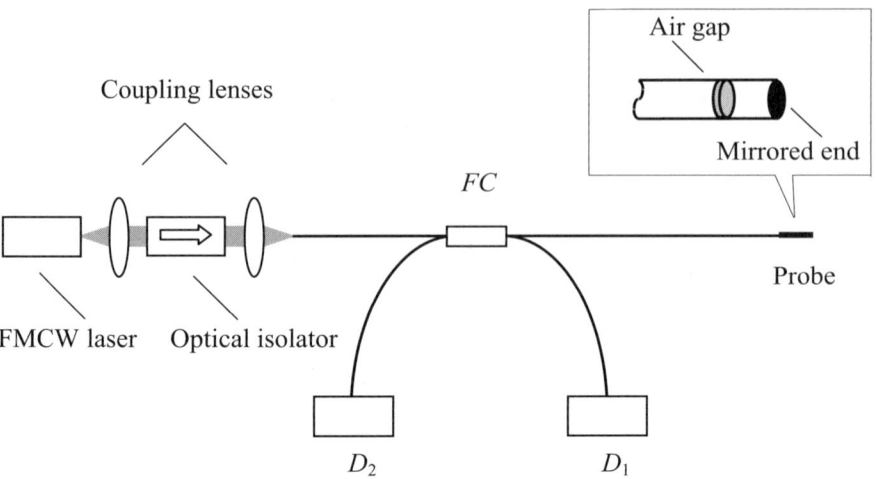

Figure 9.22 Reflectometric single-mode fiber FMCW point temperature sensor. (After [39].)

length of the single-mode fiber cavity.

If the fiber cavity is placed in a temperature field, its length will increase due to the thermal-expansion effect and its effective refractive index will change slightly. The variation of the optical path difference ΔOPD between the two reflected beams can be written as

$$\Delta OPD = 2(n_e \Delta l + l \Delta n_e)$$

$$= 2l \left[\frac{n_e \delta l}{l \delta T} + \frac{\delta n_e}{\delta T} \right] \Delta T \quad , \tag{9-81}$$

where n_e is the effective refractive index of the single-mode fiber, l is the length of the fiber cavity, Δl is the elongation of the fiber cavity, Δn_e is the variation of the effective refractive index of the fiber, ΔT is the variation of the surrounding temperature, $\delta l / l \delta T$ is the thermal expansion coefficient, and $\delta n_e / \delta T$ is the thermal dispersion coefficient (for the single-mode fiber with the core index $n = 1.456$, $\delta l / l \delta T = 5 \times 10^{-7}$ $(^\circ C)^{-1}$, $\delta n_e / \delta T = 1 \times 10^{-7}$ $(^\circ C)^{-1}$).

In the case of sawtooth-wave modulation, for instance, the intensity of the beat signal in a modulation period $I(t)$ can be written as

$$I(OPD, t) = I_0 \left[1 + V \cos \left(\frac{2\pi \Delta v v_m OPD}{c} t + \frac{2\pi}{\lambda_0} OPD \right) \right]$$

$$= I_0 [1 + V \cos(2\pi v_b t + \phi_{b0})] \quad , \tag{2-63}$$

where I_0 is the average intensity of the beat signal, V is the contrast of the beat signal, Δv is the optical frequency modulation excursion, v_m is the modulation frequency, c is the speed of light in free space, λ_0 is the central optical wavelength in free space, v_b and ϕ_{b0} are the frequency and initial phase of the beat signal, respectively. Considering Equation (9-81), the variation of the initial phase of the beat signal $\Delta\phi_{b0}$ can be written as

$$\Delta\phi_{b0} = 4\pi l \left[\frac{n_e \delta l}{l \delta T} + \frac{\delta n_e}{\delta T} \right] \frac{\Delta T}{\lambda_0}$$

(9-82)

Therefore, the temperature variation ΔT can be determined by

$$\Delta T = \frac{\lambda_0}{4\pi l} \left[\frac{n_e \delta l}{l \delta T} + \frac{\delta n_e}{\delta T} \right]^{-1} \Delta\phi_{b0}$$

(9-83)

9.5.2 Multiplexed Reflectometric Single-Mode Fiber FMCW Temperature Sensors

The reflectometric single-mode fiber FMCW interferometric temperature sensor discussed previously can be modified as a multiplexed reflectometric single-mode fiber FMCW interferometric temperature sensor by using the frequency-division multiplexing method, as shown in Figure 9.23 [39].

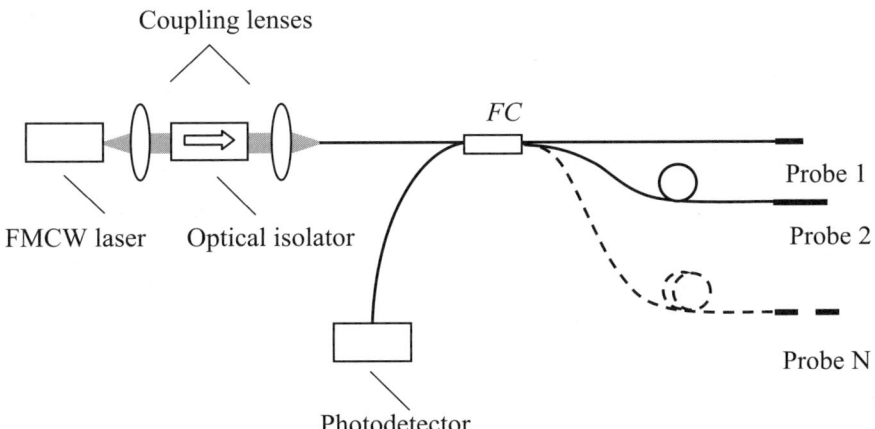

Figure 9.23. Multiplexed reflectometric single-mode fiber FMCW point temperature sensor. (After [39].)

An FMCW laser beam is launched into an input fiber of a 2×N single-mode fiber coupler (*FC*). The output fibers of the fiber coupler are connected with different lengths of fiber cavities, so that the beat signals from the different sensing probes are different in frequency and can therefore be separated by using an electric circuit with band-pass filters. Measuring the phase shift of each beat signal, we can determine the variation of the temperature around each individual sensing probe.

9.6 Fiber-optic FMCW Interferometric Rotation Sensors (Gyroscopes)

Fiber-optic interferometric rotation sensors (usually called fiber-optic gyroscopes) are used for measuring the rotation velocity of an object. Fiber-optic gyroscopes are probably the most successful fiber-optic sensors so far. Some of them have been commercialized and widely used in the areas of airplane navigation, missile guidance, instrument stabilization, and industrial robotics. Compared with mechanical gyroscopes, fiber-optic gyroscopes have a number of advantages, such as small size, no moving parts, better reliability, and rapid initiation. Compared with laser gyroscopes, fiber-optic gyroscopes are more compact (no bulk elements), flexible, safe (no electric sparks), and have low power consumption and low cost.

In the following subsections, we will first discuss the principles of fiber-optic gyroscopes, and then introduce some fiber-optic FMCW Sagnac gyroscopes, including single-mode fiber FMCW Sagnac gyroscopes, differential single-mode fiber FMCW Sagnac gyroscopes, birefringent fiber FMCW Sagnac gyroscopes and differential birefringent fiber FMCW Sagnac gyroscopes.

9.6.1 Principles of Fiber-optic Gyroscopes

Fiber-optic gyroscopes are based on the Sagnac effect: If two beams, derived from the same light source and passed through identical or almost identical paths in opposite directions, are recombined, they will interfere and construct a stable fringe. On the other hand, if the plane containing the interfering beams is rotated about its vertical axis, a phase shift in the fringe $\delta\phi$ will be introduced, which is given by

$$\delta\phi = \frac{8\pi A\Omega}{c\lambda_0},$$

(9-84)

where A is the area enclosed by the beams, Ω is the rotation angular velocity,

c is the speed of light in free space, and λ_0 is the wavelength of the light in free space. In other words, if the plane containing the interfering beams is rotated about its vertical axis, an additional optical path difference δOPD between the two interfering beams will be introduced, which is given by

$$\delta OPD = \frac{4A\Omega}{c}.$$

(9-85)

Figure 9.24(a) shows a classical Sagnac interferometer, which consists of a beam splitter (BS) and two mirrors (M_1 and M_2). The incident collimated beam is divided into two by BS. One beam goes down and is reflected by M_1 and M_2 successively, while the other beam goes up and is reflected by M_2 and M_1 successively. These two reflected beams meet at the same beam splitter BS, and parts of them propagate diagonally downward to the left to interfere at infinity. Because the two interfering beams traverse the same closed path in opposite directions, the interferometer is extremely stable and easy to align, even with an extended broadband light source. Figure 9.24(b) shows another version of the Sagnac interferometer, which uses one more mirror M_3

Figure 9.25 shows a simple single-mode fiber gyroscope in which a single-mode optical fiber coil replaces the bared air path in the Sagnac interferometer [3]. One of the important advantages of the fiber-optic gyroscope is that the Sagnac effect can be largely intensified simply by increasing the number of fiber loops, which is typically 10^3–10^4. Under this situation, the Sagnac phase shift $\delta\phi$ will be

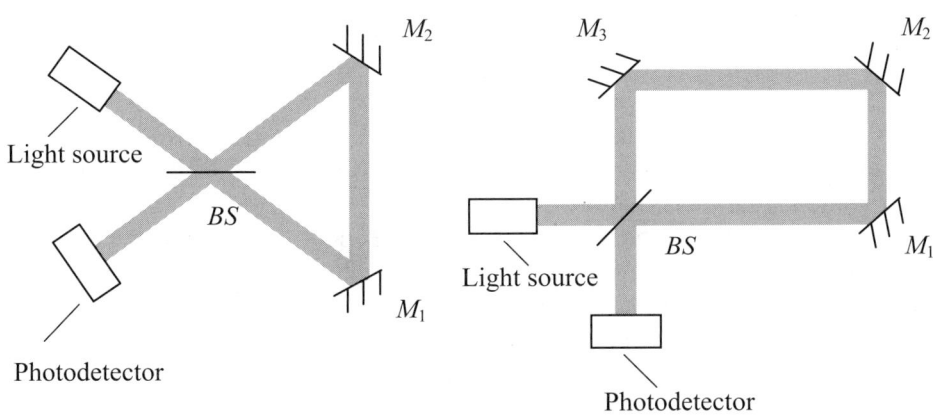

(a) A classical Sagnac interferometer.   (b) Another version of the Sagnac interferometer.

Figure 9.24. Sagnac interferometers.

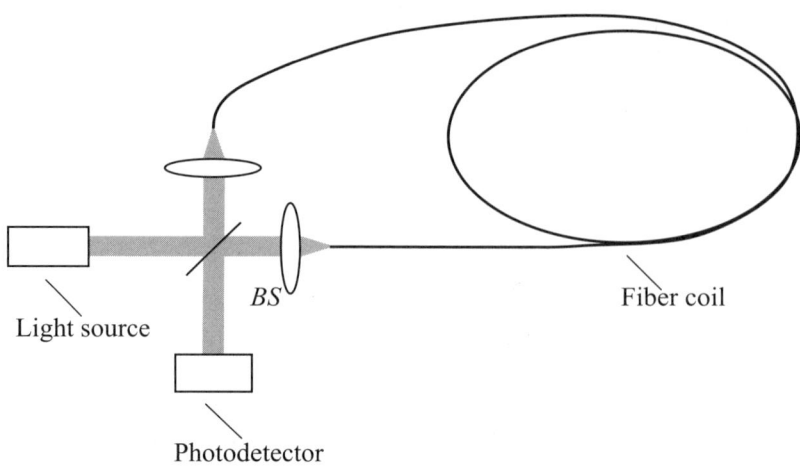

Figure 9.25. A simple single-mode fiber gyroscope. (After [3].)

$$\delta\phi = \frac{8\pi AN\Omega}{c\lambda_0},$$

(9-86)

where N is the number of fiber loops. The corresponding optical path differ-ence change δOPD will be

$$\delta OPD = \frac{4AN\Omega}{c}.$$

(9-87)

Figure 9.26 shows a simple single-mode all-fiber gyroscope, in which a fi-ber-optic coupler (FC) is used to replace the bulk beam splitter and some bared air paths in the previous system [11]. Obviously, the all-fiber gyroscope is more compact, flexible, and reliable.

Finding the Sagnac phase shift requires the theory of general relativity be-cause it involves light propagating in an accelerated system. However, we still can discuss it in the inertial system and assume that the propagation of light is independent on the motion of the medium [13].

Now, let's assume that the fiber coil is circular, as shown in Figure 9.27. When the fiber coil is rotating around the vertical axis at an angular velocity Ω, an observer at rest in the inertial frame of reference sees that the launching point M will move to M'. The length L_1 passed by the clockwise propagating beam through the whole fiber coil is

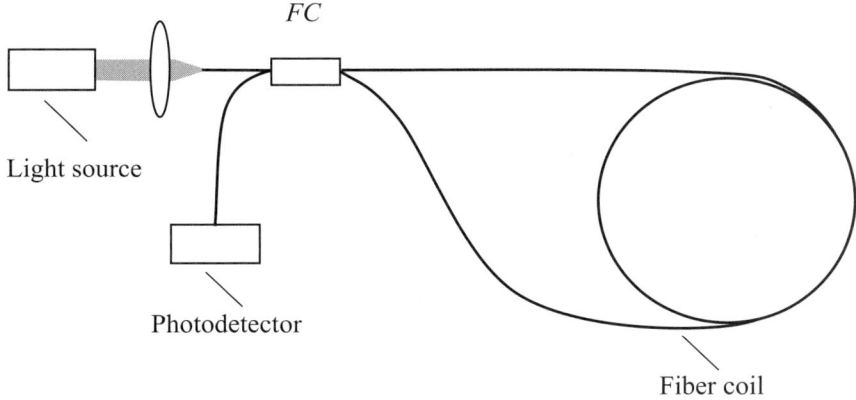

Figure 9.26. A simple single-mode all-fiber gyroscope. (After [11].)

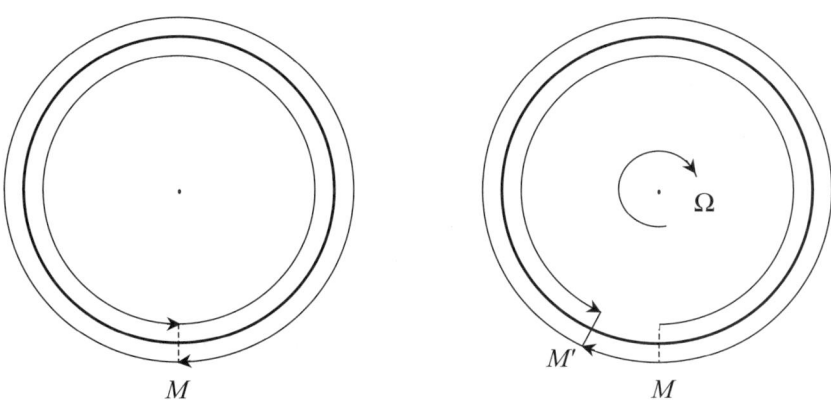

(a) At rest in an inertial frame of reference.

(b) Rotating with respect to an inertial frame of reference.

Figure 9.27. Sagnac effect in a single-mode fiber coil.

$$L_1 = l + \Delta l$$
$$= l + R\Omega T ,$$
$$(9\text{-}88)$$

where Δl is the fiber length between the points M and M', R is the radius of the fiber coil, and T is the average period when the beams propagate through the whole fiber coil, given by

$$T = \frac{2\pi R}{c} .$$

$$\text{(9-89)}$$

Similarly, the length L_2 passed by the anticlockwise propagating beam is

$$L_2 = l - \Delta l$$
$$= l - R\Omega T .$$

$$\text{(9-90)}$$

Therefore, the additional optical path difference δOPD between the two counterpropagating beams equals

$$\delta OPD = L_1 - L_2$$
$$= 2R\Omega T$$
$$= \frac{4A\Omega}{c} ,$$

$$\text{(9-85)}$$

where A is the area enclosed by the fiber coil ($A = \pi R^2$). Correspondingly, the equivalent phase shift in the interferential signal is

$$\delta\phi = \frac{8\pi A\Omega}{c\lambda_0} ,$$

$$\text{(9-84)}$$

where λ_0 is the light wavelength in free space.

If the fiber coil has many loops, the additional optical path difference due to rotation will be

$$\delta OPD = \frac{4AN\Omega}{c} ,$$

$$\text{(9-87)}$$

where N is the number of fiber loops, or

$$\delta OPD = \frac{2RL\Omega}{c} ,$$

$$\text{(9-91)}$$

where L is the total length of the fiber coil. Correspondingly, the Sagnac phase shift will be

$$\delta\phi = \frac{8\pi AN\Omega}{c\lambda_0}$$

$$\text{(9-86)}$$

or

$$\delta\phi = \frac{4\pi RL\Omega}{c\lambda_0}.$$

(9-92)

Usually, the relationship between the rotation velocity and the Sagnac phase shift is represented by

$$\Omega = k\delta\phi,$$

(9-93)

where k is named as the scale factor in $\sec^{-1}$. Obviously, the scale factor of the fiber-optic gyroscope equals

$$k = \frac{c\lambda_0}{8\pi AN}$$
$$= \frac{c\lambda_0}{4\pi RL}.$$

(9-94)

In general, fiber-optic gyroscopes are required to have a high level of sensitivity. For instance, assuming that a fiber-optic gyroscope has a single-mode fiber coil of 0.1 meter in radius and 500 meters in length (about 800 loops), and assuming that the laser operates at 0.660 μm, the scale factor k equals

$$k = 0.32.$$

(9-95

If the fiber-optic gyroscope can measure Earth's rotation speed (4.2×10^{-3} deg/s), the resolution of the phase shift should be at least

$$(\delta\phi)_{min} = 1.3\times10^{-2}(deg).$$

(9-96)

Increasing the length of the fiber coil, the radius of the fiber coil, or both can improve the sensitivity of the fiber-optic gyroscope. However, the total length of the fiber coil in a real fiber-optic gyroscope is limited by energy loss, and the radius of the fiber coil is limited by other constraints. Therefore, other possible approaches to improving the sensitivity of the fiber-optic gyroscope have to be found.

One approach to improving the sensitivity is changing the operating point of the fiber-optic gyroscope. In general, the output intensity from the fiber coil can be written as

$$I = I_0(1 + V\cos\delta\phi),$$

(9-97)

where I_0 is the average intensity, V is the contrast of the signal, and $\delta\phi$ is the Sagnac phase shift, as shown in Figure 9.28(a). The sensitivity of the fiber-optic gyroscope usually is represented by the derivative of the intensity of the output signal,

$$\frac{dI}{d(\delta\phi)} = -I_0 V \sin \delta\phi$$

(9-98)

Apparently, the ordinary gyroscope is working on the zero-sensitivity point where the rotation velocity of the gyroscope is approaching zero ($\delta\phi = 0$, $dI/d(\delta\phi) = 0$).

For maximum sensitivity, the operating point should be shifted to the quadrature point, as shown in Figure 9.28(b), where, with no rotation, the phase difference between the two interfering beams is $\pi/2$. The output intensity from the phase-shifted gyroscope can be written as

$$I = I_0 \left[1 + V \cos\left(\delta\phi + \frac{\pi}{2} \right) \right]$$

(9-99)

and the derivative of the output intensity is

$$\frac{dI}{d(\delta\phi)} = -I_0 V \sin\left(\delta\phi + \frac{\pi}{2} \right)$$
$$= -I_0 V \cos \delta\phi$$

(9-100)

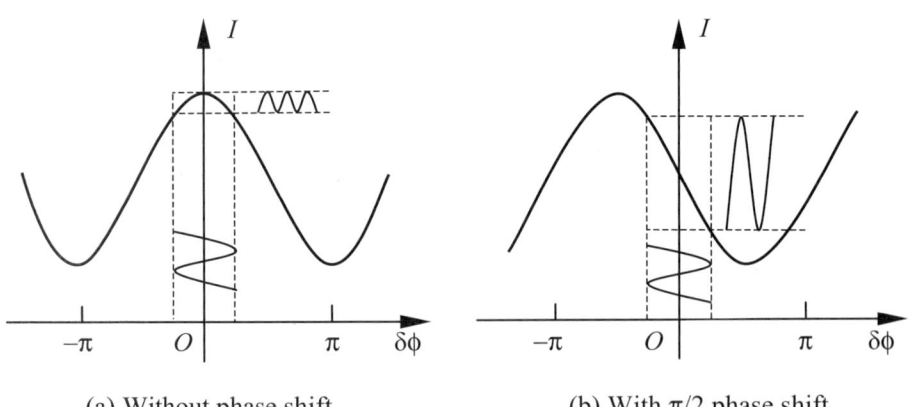

(a) Without phase shift.   (b) With $\pi/2$ phase shift.

Figure 9.28. Operating points of fiber-optic gyroscopes.

For small $\delta\phi$, the equations above can be simplified as

$$I \approx I_0(1 - V\delta\phi) \,,$$

(9-101)

$$\frac{dI}{d(\delta\phi)} \approx -I_0 V .$$

(9-102)

Obviously, when the rotation velocity is approaching zero, the sensitivity of the phase-shifted gyroscope equals the product of the average intensity and contrast of the signal rather than zero.

The limitations of this approach include: (i) the supplementary component for the $\pi/2$ phase shift must be stable; or it will introduce an unexpected nonreciprocal phase drift in the real environment; (ii) because the output intensity is a static signal (i.e., the Sagnac phase shift is determined by measuring the variation of the signal intensity), any erratic reflection or loss in the optical system or intensity noise of the light source will introduce a measurement error and thus limit the accuracy of the fiber-optic gyroscope; (iii) the dynamic range of this type of fiber-optic gyroscope generally is limited to $\pm\pi/2$.

Another approach to improving the sensitivity is to dynamically modulate the phase difference of the two counterpropagating beams. Figure 9.29 shows a single-mode all-fiber phase-modulated gyroscope in which a phase modulator consisting of a few turns of fiber wound around a piezoelectric cylinder is

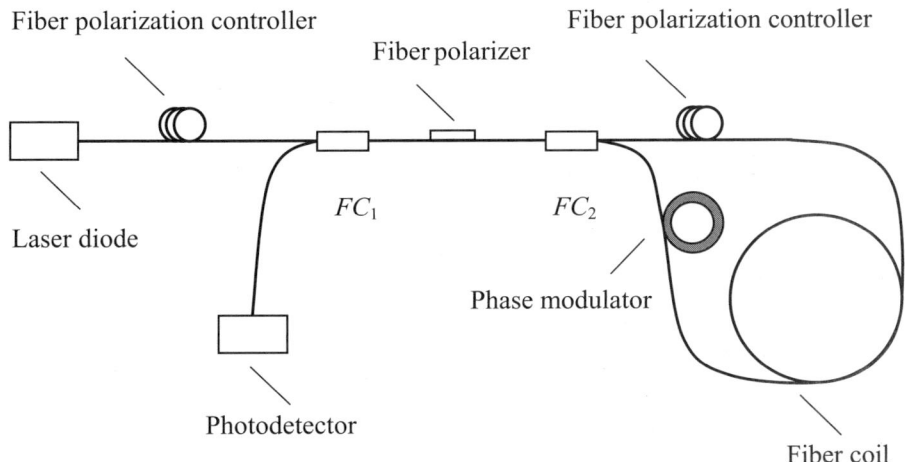

Figure 9. 29. Single-mode all-fiber phase-modulated gyroscope. (After [12].)

located near one end of the optical fiber coil [12]. A laser diode with an optical fiber pigtail is employed as the light source so that the stability of the fiber-optic gyroscope can be further improved due to the nature of the solid interferometer structure. A fiber-optic polarizer and two in-line fiber-optic polarization controllers are employed to ensure that the two counterpropagating beams travel in the same polarization mode, and thus the maximum signal contrast can be obtained.

The piezoelectric cylinder alters its diameter in sympathy with the outside electric signal. The stress produced modulates the length and the effective refractive index of the fiber wound around a piezoelectric cylinder and in turn the phase of the guided wave. Because of reciprocity, the amplitude of phase modulation is exactly the same for both clockwise and anticlockwise propagating beams. However, a relative phase difference is produced between the two beams due to the propagation time through the fiber. The output signal generally is preprocessed with a phase-locked amplifier, and the second harmonic is used for the calibration of the rotation velocity.

Figure 9.30 shows another version of the single-mode all-fiber phase-modulated gyroscope, in which a Y-type waveguide coupler and two metal electrode phase modulators are fabricated on a LiNbO₃ substrate material by using integrated optics (IO) technology [31]. The advantages of this configuration include that it can work in a closed-loop mode, where one of the phase modulators is used for the phase modulation while another is used for producing a reverse phase shift to cancel out the phase shift due to the rotation, and that the IO component can function as a high-quality polarizer.

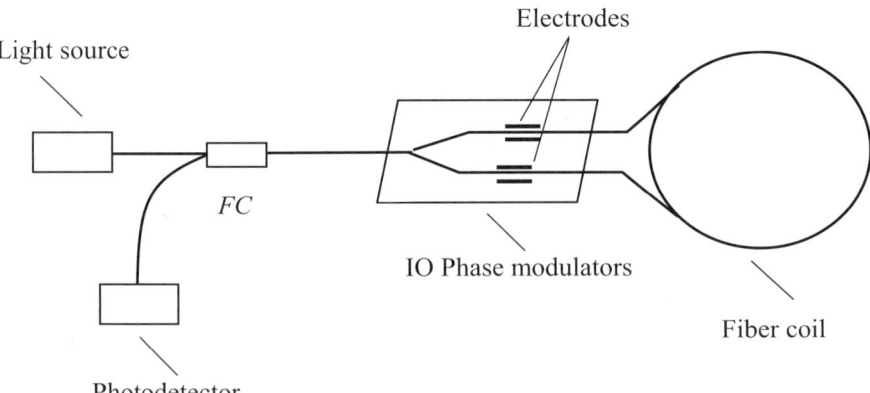

Figure 9.30. Single-mode all-fiber phase-modulated gyroscope with an IO component. (After [31].)

Frequency-modulated continuous-wave interference naturally produces a dynamic signal and therefore has no problem of a zero-sensitivity point. The potential problem in a fiber-optic FMCW gyroscope is that, due to the nature of the unbalanced interferometer structure, there may be an unexpected nonreciprocal phase drift if the environmental conditions are unstable. This nonreciprocal phase drift can significantly affect the accuracy and the long-term stability of the gyroscope. However, if the fiber-optic network is properly designed, this nonreciprocal phase drift can be compensated optically or electrically. Considering other advantages of FMCW interference, such as high accuracy (easy calibration of the fractional phase and immunity to the light power variation), large measurement range (no $\pm\pi/2$ phase-shift restriction and no ambiguous fringe-count problem), simple signal processing (quasi-sinusoidal beat signal), compact size, and light weight (no bulk phase modulator), frequency-modulated continuous-wave interference technology could be a new and efficient way to develop advanced fiber-optic gyroscopes.

9.6.2 Single-Mode Fiber FMCW Gyroscopes

Figure 9.31 shows schematically a single-mode fiber FMCW gyroscope, which consists of a laser diode, two polarizers (P_1 and P_2), two coupling lenses, four Y-type single-mode fiber-optic couplers (FC_1–FC_4), a single-mode fiber coil, a photodetector, and an in-line fiber polarization controller (PC) [105]. The two output fibers of FC_1 and FC_3 are connected with the two input fibers of FC_2 and FC_4 to construct the basic structure of an unbalanced interferometer. The output fibers of FC_2 and FC_4 are connected with the fiber coil. The two polarizers and the in-line fiber polarization controller are used to ensure that the two counterpropagating beams travel in the same polarization mode, and thus the contrast of the beat signal can be optimized.

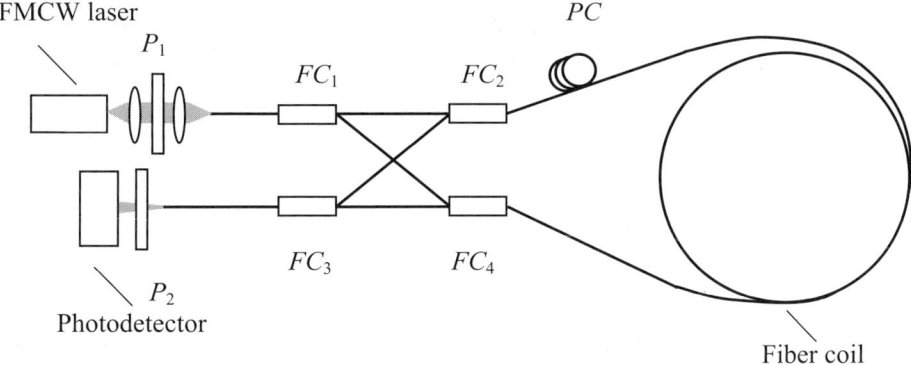

Figure 9.31. Single-mode fiber FMCW gyroscope. (After [105].)

A polarized FMCW laser beam is first divided into two beams by FC_1. One passes through FC_2 and enters the upper end of the fiber coil, while another passes through FC_4 and enters the lower end of the fiber coil. These two beams propagate in opposite directions, again separately pass through the couplers FC_4 and FC_2, and enter the fiber coupler FC_3 to mix coherently. The beat signal produced exits from FC_3 and finally is received by a photodetector.

The initial optical path difference between the two counter-propagating beams in the fiber coil can be written as

$$OPD = n_e (l_{12} + l_{34} - l_{14} - l_{23}),$$ (9-103)

where n_e is the effective refractive index of the single-mode fiber, l_{12}, l_{34}, l_{14}, and l_{23} are the lengths of the linking fibers from FC_1 to FC_2, FC_3 to FC_4, FC_1 to FC_4, and FC_2 to FC_3, respectively. Properly choosing the lengths of these fibers, we can get any desired initial optical path difference.

If the fiber coil is in rotation, the additional optical path difference δOPD due to the Sagnac effect can be written as

$$\delta OPD = \frac{4AN\Omega}{c},$$ (9-87)

where A is the area enclosed by the fiber coil, N is the number of fiber loops, Ω is the rotation angular velocity component parallel to the coil axis, and c is the speed of light in free space, or written as

$$\delta OPD = \frac{2RL\Omega}{c},$$ (9-91)

where R is the radius of the fiber coil and L is the total length of the fiber coil.

If the laser frequency is modulated with a sawtooth waveform, the intensity of the detected beat signal in a modulation period $I(t)$ would be

$$I(OPD,t) = I_0 \left[1 + V \cos \left(\frac{2\pi \Delta \nu \nu_m OPD}{c} t + \frac{2\pi}{\lambda_0} OPD \right) \right]$$
$$= I_0 [1 + V \cos(2\pi \nu_b t + \phi_{b0})],$$ (2-63)

where I_0 is the average intensity of the beat signal, V is the contrast of the beat signal, $\Delta \nu$ is the optical frequency modulation excursion, ν_m is the modulation frequency, c is the speed of light in free space, λ_0 is the central optical wavelength in free space, and ν_b and ϕ_{b0} are the frequency and initial phase of the beat signal, respectively. Considering Equation (9-91), the Sagnac phase shift $\delta\phi_{b0}$ can be written as

$$\delta\phi_{b0} = \frac{8\pi A N \Omega}{c\lambda_0}$$

$$= \frac{4\pi R L \Omega}{c\lambda_0} . \tag{9-104}$$

where λ_0 is the central wavelength in free space. Therefore, the rotation angular velocity Ω of the fiber coil can be determined by

$$\Omega = \frac{c\lambda_0}{8\pi A N} \delta\phi_{b0}$$

$$= \frac{c\lambda_0}{4\pi R L} \delta\phi_{b0}$$

$$= k\delta\phi_{b0} , \tag{9-105}$$

where k is the scale factor of the single-mode fiber FMCW gyroscope, given by

$$k = \frac{c\lambda_0}{8\pi A N}$$

$$= \frac{c\lambda_0}{4\pi R L} . \tag{9-106}$$

The advantages of the single-mode fiber FMCW gyroscope include high resolution, large measurement range, and simple signal processing. The limitation is that the linking fibers, which are necessary for introducing an initial optical path difference in the gyroscope, may introduce an unexpected nonreciprocal phase drift if the environmental conditions (such as temperature) change, and therefore a proper temperature-control system should be used in practice.

9.6.3 Differential Single-Mode Fiber FMCW Gyroscopes

The configuration of the differential single-mode fiber FMCW gyroscope is similar to that of the single-mode fiber FMCW gyroscope. The difference is that the two Y-type fiber couplers, FC_2 and FC_4, in the previous gyroscope are replaced by two X-type fiber couplers whose additional output fibers are directly connected to each other to form a shortcut and that the laser is driven by a gated modulation signal, as shown in Figure 9.32.

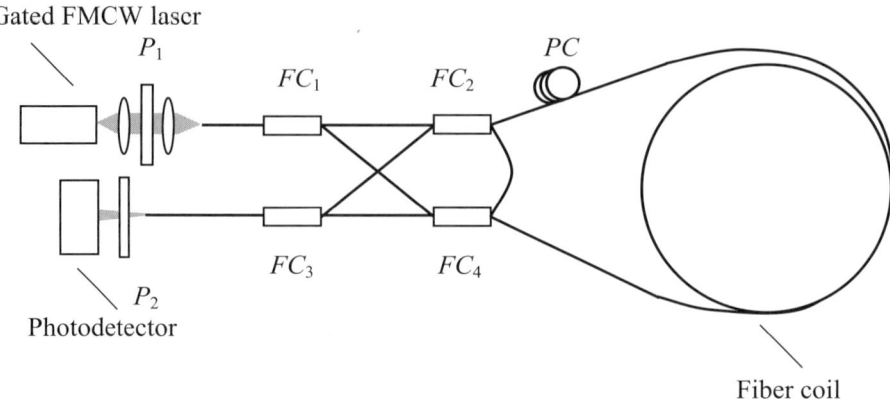

Figure 9.32. Differential single-mode fiber FMCW gyroscope.

The differential single-mode fiber FMCW gyroscope actually is a time-division multiplexed single-mode fiber FMCW gyroscope. A gated FMCW laser beam is first launched into the fiber coupler FC_1 and split into two beams. When these beams pass through the couplers FC_2 and FC_4, they will split again. Parts of the two beams traverse the fiber coil in opposite directions and parts of the two beams take the shortcut in opposite directions. According to the conclusion in Section 8.2, if the delay time introduced by the fiber coil is longer than the frequency modulation period but shorter than the separation of the gated wavelet, the beat signal produced by the two counterpropagating beams in the fiber coil and the beat signal produced by the two counterpropagating beams in the shortcut will arrive on the photodetector at different times without overlapping and can therefore be separated with an electric gate circuit.

Obviously, both beat signals contain the same nonreciprocal phase drift caused by variation of the environmental conditions (including the frequency drift of the laser), but only the beat signal from the fiber coil contains the reciprocal Sagnac phase shift if the fiber coil is in rotation. Therefore, separating the two beat signals with an electric gate circuit and finding the phase difference between them, we can discover the Sagnac phase shift caused purely by the rotation of the fiber coil.

9.6.4 Birefringent Fiber FMCW Gyroscopes

We know that a single length of birefringent fiber can be used as a two-beam unbalanced interferometer if the light is coupled into the two orthogonal polarization modes. Therefore, for a birefringent fiber FMCW gyroscope, it is not necessary to build an extra fiber network for introducing an initial optical path difference.

Figure 9.33 shows a birefringent fiber FMCW gyroscope, which consists of a laser diode, a birefringent fiber coil, a polarizer (*P*), a beam splitter (*BS*), three collimating and coupling lenses, and a photodetector [78]. A polarized FMCW laser beam is first divided into two beams by *BS*. These two beams are then coupled into the birefringent fiber coil in different polarization modes from the different ends. For instance, the clockwise propagating beam is coupled into the $HE_{11}{}^x$ mode, while the anticlockwise propagating beam is coupled into the $HE_{11}{}^y$ mode. After traversing the fiber coil, the two beams are recombined by the same beam splitter *BS*. The beat signal produced is finally detected by the photodetector.

Notice that, because the polarization direction of each beam rotates 90° after emerging from the birefringent fiber, the polarization directions of the two exiting beams are still parallel to each other, but they are perpendicular to the polarization direction of the incident beams. This property of the birefringent fiber gyroscope is very important because it ensures that the best signal contrast is always obtained.

The optical path difference *OPD* between the two counterpropagating beams in the birefringent fiber coil can be written as

$$OPD = (n_x - n_y)l,$$

(9-107)

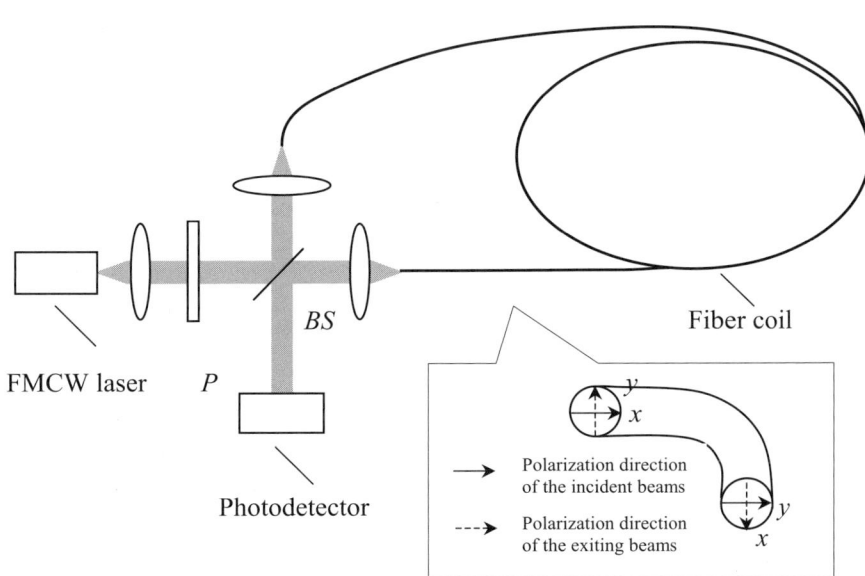

Figure 9.33. Birefringent fiber FMCW gyroscope. (After [78].)

where n_x is the effective refractive index of the $HE_{11}{}^x$ mode, n_y is the effective refractive index of the $HE_{11}{}^y$ mode, and l is the length of the birefringent fiber coil. If the laser frequency is modulated with a sawtooth waveform, the intensity of the detected beat signal in a modulation period $I(t)$ can be written as

$$I(OPD,t) = I_0 \left[1 + V \cos \left(\frac{2\pi \Delta v v_m OPD}{c} t + \frac{2\pi}{\lambda_0} OPD \right) \right]$$

$$= I_0 [1 + V \cos(2\pi v_b t + \phi_{b0})] \qquad , \qquad (2\text{-}63)$$

where I_0 is the average intensity of the beat signal, V is the contrast of the beat signal, Δv is the optical frequency modulation excursion, v_m is the modulation frequency, c is the speed of light in free space, λ_0 is the central optical wavelength in free space, and v_b and ϕ_{b0} are the frequency and initial phase of the beat signal respectively.

The Sagnac phase shift $\delta\phi_{b0}$ in the beat signal can be written as

$$\delta\phi_{b0} = \frac{8\pi A N \Omega}{c\lambda_0}$$

$$= \frac{4\pi R L \Omega}{c\lambda_0} \qquad , \qquad (9\text{-}104)$$

where λ_0 is the central wavelength in free space. Therefore, Ω is equal to

$$\Omega = \frac{c\lambda_0}{8\pi A N} \delta\phi_{b0}$$

$$= \frac{c\lambda_0}{4\pi R L} \delta\phi_{b0}$$

$$= k\delta\phi_{b0} \qquad , \qquad (9\text{-}105)$$

where k is the scale factor of the birefringent fiber FMCW gyroscope given by

$$k = \frac{c\lambda_0}{8\pi A N}$$

$$= \frac{c\lambda_0}{4\pi R L} \qquad . \qquad (9\text{-}106)$$

The advantages of the birefringent fiber FMCW gyroscope include simple configuration and no polarization noise (for the ideal birefringent fiber). The drawback is that the nonreciprocal optical path difference in the birefringent fiber coil can introduce an unexpected nonreciprocal phase drift if the environmental parameters change. On the other hand, this phenomenon offers the possibility of measuring environmental parameters, such as strain and temperature, with the birefringent fiber Sagnac interferometer.

9.6.5 Differential Birefringent Fiber FMCW Gyroscopes

The differential birefringent fiber FMCW gyroscope is actually a polarization-division multiplexed birefringent fiber FMCW gyroscope, as shown in Figure 9.34. Compared with the previous birefringent fiber FMCW gyroscope, the major difference of the differential one is that the two beams launching into the birefringent fiber coil are coupled into both the HE_{11}^x mode and the HE_{11}^y mode. (This can be realized simply by rotating the polarizer (P) 45°.) Therefore, there will be two clockwise propagating beams and two anticlockwise propagating beams in the birefringent fiber. In other words, there will be two HE_{11}^x mode beams and two HE_{11}^y mode beams propagating in the birefringent fiber.

Note that, similar to the case of the previous birefringent fiber gyroscope, the coordinates of the principal axes on the two ends of the birefringent fiber coil have a 90° rotation and thus the clockwise propagating HE_{11}^x mode beam and anticlockwise propagating HE_{11}^y mode beam will vibrate in the same direction after exiting the birefringent fiber coil and produce a beat signal, while the

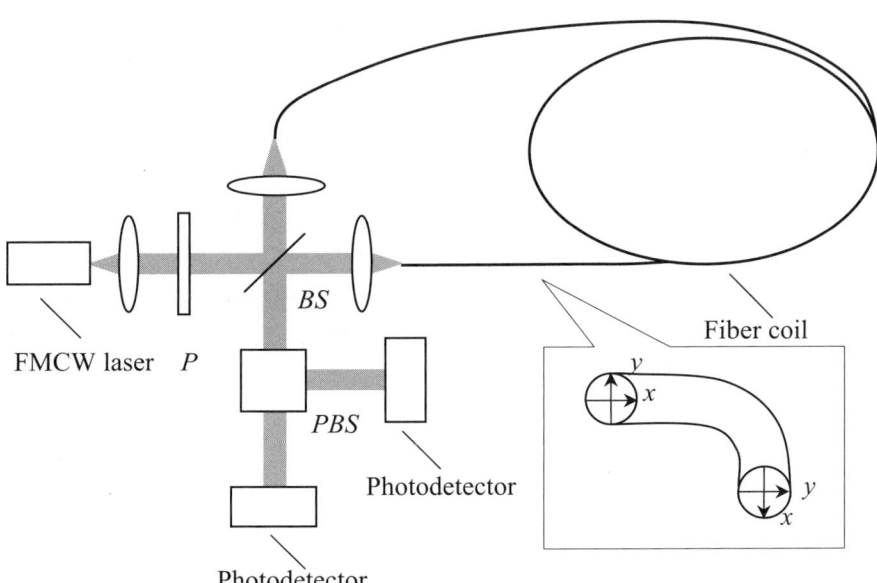

Figure 9.34. Differential birefringent fiber FMCW gyroscope.

clockwise propagating $HE_{11}{}^y$ mode beam and anticlockwise propagating $HE_{11}{}^x$ mode beam will vibrate in another orthogonal direction after exiting the bire-fringent fiber coil and produce another beat signal. These two beat signals are naturally perpendicular to each other, so they can be separated by using a po-larization beam splitter (*PBS*). The separated beat signals are finally detected by two photodetectors.

One of the important properties of this configuration is that the two beat signals contain the same initial optical path difference but an opposite recipro-cal Sagnac variation of optical path difference if the fiber coil is in rotation. For instance, assuming $n_{ex} > n_{ey}$, if we take the short optical path as the reference for both beat signals, the initial optical path differences OPD_1 and OPD_2 equal

$$OPD_1 = OPD_2 = (n_{ex} - n_{ey})L , \qquad (9\text{-}107)$$

where n_{ex} and n_{ey} are the effective refractive indexes of the $HE_{11}{}^x$ mode and the $HE_{11}{}^y$ mode respectively, L is the total length of the birefringent fiber coil. When the fiber coil rotates clockwise, the long optical path for the clockwise propagating $HE_{11}{}^x$ mode beam gets longer, while the short optical path for the anticlockwise propagating $HE_{11}{}^y$ mode beam gets shorter, so that the optical path difference in the first beat signal increases, as shown in Figure 9.35(a). Concurrently, the short optical path for the clockwise propagating $HE_{11}{}^y$ mode beam gets longer, while the long optical path for the anticlockwise propagating $HE_{11}{}^x$ mode beam gets shorter, so that the optical path difference in the second beat signal decreases, as shown in Figure 9.35(b).

In other words, when the fiber coil rotates clockwise, the Sagnac variation of optical path difference δOPD_1 produced by the clockwise propagating $HE_{11}{}^x$ mode beam and anticlockwise propagating $HE_{11}{}^y$ mode beam will be

$$\delta OPD_1 = \frac{2RL\Omega}{c} . \qquad (9\text{-}108)$$

On the other hand, the Sagnac variation of optical path difference δOPD_2 pro-duced by the clockwise propagating $HE_{11}{}^y$ mode beam and anticlockwise propagating $HE_{11}{}^x$ mode beam will be

$$\delta OPD_2 = \frac{-2RL\Omega}{c} . \qquad (9\text{-}109)$$

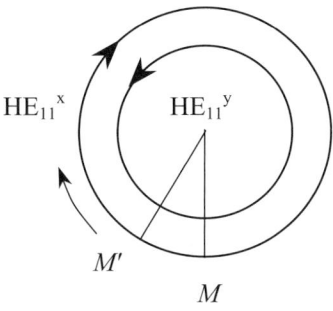 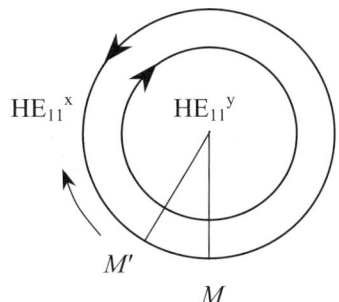

(a) OPD variation of the clockwise propagating $HE_{11}{}^x$ mode beam and the anticlockwise propagating $HE_{11}{}^y$ mode beam.

(b) OPD variation of the clockwise propagating $HE_{11}{}^y$ mode beam and the anticlockwise propagating $HE_{11}{}^x$ mode beam.

Figure 9.35. Sagnac effect on a birefringent fiber coil.

Therefore, the Sagnac phase shifts of the two beat signals, $\delta\phi_{b01}$ and $\delta\phi_{b02}$, will be

$$\delta\phi_{b01} = \frac{4\pi RL\Omega}{c\lambda_0} , \tag{9-110}$$

$$\delta\phi_{b02} = \frac{-4\pi RL\Omega}{c\lambda_0} . \tag{9-111}$$

The phase difference of the two beat signals $\Delta(\delta\phi_{b0})$ equals

$$\Delta(\delta\phi_{b0}) = \delta\phi_{b01} - \delta\phi_{b02}$$
$$= \frac{8\pi RL\Omega}{c\lambda_0} . \tag{9-112}$$

The rotation angular velocity Ω can be determined by

$$\Omega = \frac{c\lambda_0}{8\pi RL}\Delta(\delta\phi_{b0})$$
$$= k\Delta(\delta\phi_{b0}) , \tag{9-113}$$

where k is the scale factor of the differential birefringent fiber FMCW gyro-scope, given by

$$k = \frac{c\lambda_0}{8\pi RL}.$$

(9-114)

Obviously, the differential birefringent fiber FMCW gyroscope not only over-comes the problem of nonreciprocal phase drift, but also doubles the rotational sensitivity.

Differential fiber-optic FMCW gyroscopes (including the differential single-mode fiber gyroscope and the differential birefringent fiber gyroscope) are immune from the effects of the environmental conditions and the frequency drift of the laser. This is extremely important because it significantly improves the accuracy and the long-term stability of the gyroscopes. Considering their other advantages, such as high resolution, large measurement range, simple signal, full passiveness, compact size, and light weight, differential fiber-optic FMCW gyroscopes will receive more and more interest.

Chapter 10

Signal Processing of Optical Frequency-Modulated Continuous-Wave Interference

From the previous chapters, we can see that all information in an optical FMCW interferometric system is embedded in the beat signal. Both the frequency and the initial phase of the beat signal are related to the optical path difference between the two interfering optical waves in the system. The frequency of the beat signal determines the absolute value of the optical path difference (or the absolute value of the associated parameter), while the variation of the initial phase (i.e., the phase shift) of the beat signal determines the change of the optical path difference (or the change of the associated parameter). Hence, the ultimate task for any optical FMCW interferometric system is determining the frequency and the phase shift of the beat signal.

The simplest way to measure the frequency and the phase shift of a beat signal may be by using an electronic oscilloscope to analyze the waveform of the beat signal. For instance, the period of the beat signal can be determined by measuring the time interval between two neighboring peaks (or two neighboring zeroes), and the frequency of the beat signal v_b can be calculated with the equation

$$v_b = \frac{1}{T_b},$$

(10-1)

where T_b is the period of the beat signal.

For the beat signal from a sawtooth-wave optical FMCW interferometer, the phase shift can be determined by watching the movement of the signal pattern, usually using the waveform of the modulation signal as a reference. For the beat signal from a triangular-wave or sinusoidal-wave optical FMCW interferometer, because the phase-shift directions of the beat signal in the rising and falling periods are opposite, the phase shift can be determined by viewing the change in intensity at the junctions of the rising and falling periods. For instance,

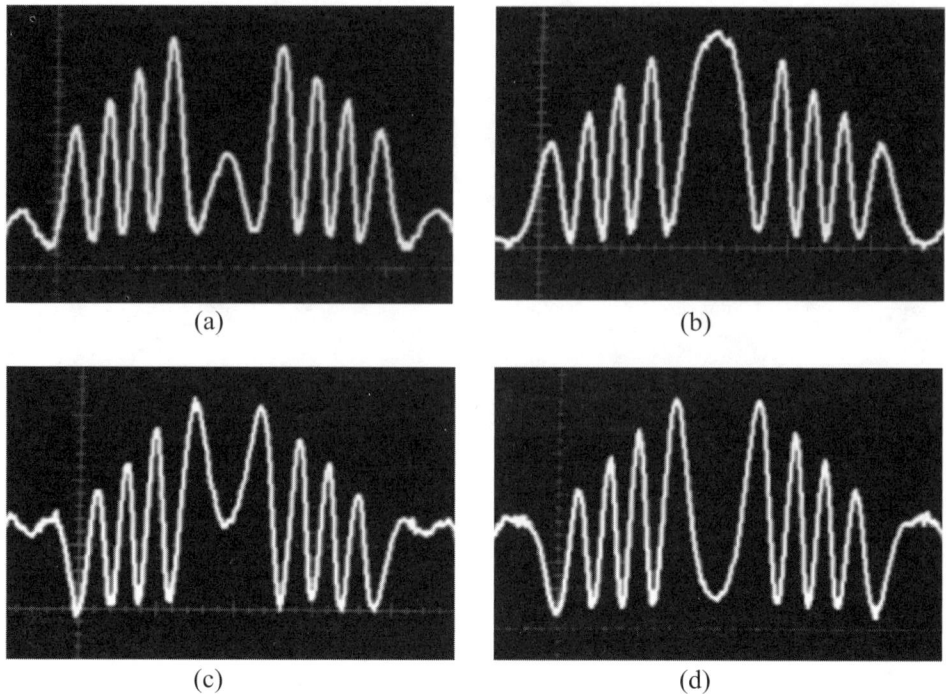

(a)   (b)

(c)   (d)

Figure 10.1. Waveforms of the beat signal from a sinusoidal-wave optical FMCW interferometer. (Photo by the author.)

with respect to the waveform in Figure 10.1(a), the waveforms in Figures 10.1(b), (c), and (d) have been shifted by $\pi/2$, π, and $3\pi/2$ radians, respectively.

Using an oscilloscope to determine the frequency and phase shift of a beat signal is visual and easy. However, for automatic signal processing or instrumentation, an electric circuit is necessary. In the following sections, we will introduce some practical electronic methods for frequency measurement and phase measurement.

10.1 Frequency Measurement

Frequency measurement is a typical topic in electrical measurement technology. Some existing frequency measurement methods, after a little modification, can still be used for frequency measurement in optical FMCW interference. In this section, we will discuss three commonly used methods: the time-fixed cycle-counting method, the number-fixed cycle-counting method, and the pulse filling method.

10.1.1 Time-Fixed Cycle-Counting Method

In this measurement, a fixed-width gate signal is used to periodically frame the beat signal, and the output signal is then counted using a pulse counter. For the beat signal from a sawtooth-wave optical FMCW interferometer, for instance, in order to avoid the error due to the phase discontinuity at the junctions of the modulation periods, the framing gate signal should have the same frequency as the modulation signal, the gate opened at the starting edge of each modulation period, and the gate width restricted to be shorter than the modulation period, as shown in Figure 10.2.

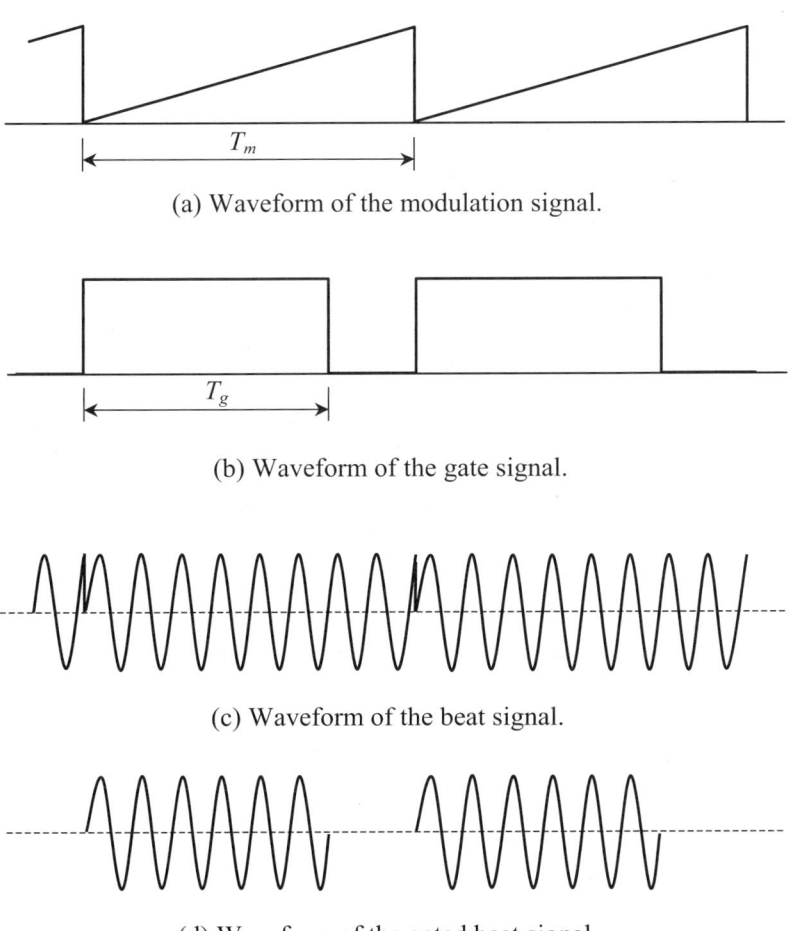

(a) Waveform of the modulation signal.

(b) Waveform of the gate signal.

(c) Waveform of the beat signal.

(d) Waveform of the gated beat signal.

Figure 10.2. Time-fixed cycle-counting method for frequency measurement.

The frequency of the beat signal v_b can be determined by

$$v_b = \frac{N}{T_g},$$

(10-2)

where N is the number of cycles of the beat signal inside the gate and T_g is the gate width. The gate width usually is quantified by using another high-frequency pulse signal.

The maximum relative error $\Delta v_b / v_b$ is determined by the fractional portion of the cycles (usually called quantification error) and the inaccuracy of the gate width, and it can be written as

$$\frac{\Delta v_b}{v_b} = \frac{\Delta N}{N} + \frac{\Delta T_g}{T_g}$$

$$= \frac{1}{N} + \frac{\Delta T_p}{T_p}$$

$$= \frac{1}{v_b T_g} + \frac{\Delta v_p}{v_p},$$

(10-3)

where T_p is the period of the pulse signal, ΔT_p is the pulse period error, v_p is the frequency of the pulse signal, and Δv_p is the pulse frequency error.

Since the maximum relative error is inversely proportional to the beat frequency, this method is suitable for measuring the beat signal of high frequency.

10.1.2 Number-Fixed Cycle-Counting Method

The number-fixed cycle-counting method still uses a gate signal to periodically frame the beat signal, but the number of cycles of the beat signal is fixed and the gate width is variable. Measuring the gate width by using another high-frequency pulse signal, we can determine the frequency of the input beat signal, as shown in Figure 10.3.

The frequency of the beat signal v_b can be determined by

$$v_b = \frac{N}{T_g}$$

$$= \frac{N}{MT_p},$$

(10-4)

(a) Waveform of the input beat signal.

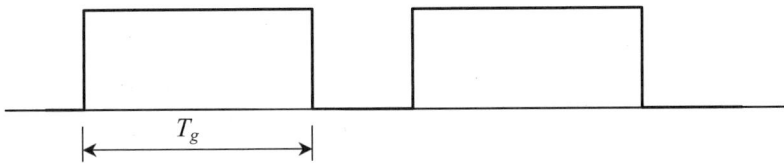

(b) Waveform of the gate signal.

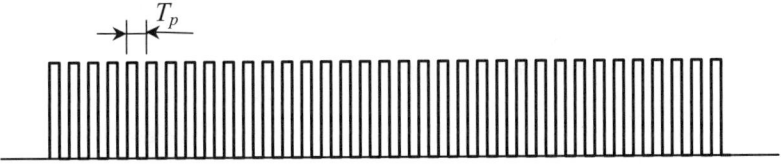

(c) Waveform of the high-frequency pulse signal.

(d) Waveform of the gated high-frequency pulse signal.

Figure 10.3. Number-fixed cycle-counting method for frequency measurement.

where N is the number of cycles of the beat signal within the gate, T_g is the gate width, M is the number of high-frequency pulses inside the gate, and T_p is the pulse period.

The maximum relative error $\Delta v_b / v_b$ can be written as

$$\frac{\Delta v_b}{v_b} = \frac{\Delta M}{M} + \frac{\Delta T_p}{T_p}$$

$$= \frac{1}{M} + \frac{\Delta v_p}{v_p}$$

$$= \frac{v_b}{Nv_p} + \frac{\Delta v_p}{v_p} , \tag{10-5}$$

where v_p is the frequency of the pulse signal and Δv_p is the pulse frequency error. The first term is the quantification error, and the second term is the error caused by the inaccuracy of the pulse period.

Obviously, the number-fixed cycle-counting method is suitable for the low-frequency beat signal. However, because the gate width is required to be shorter than the modulation period, the frequency of the beat signal must be higher than N/T_m (where T_m is the modulation period of the laser).

The cycle-counting methods (including the time-fixed cycle-counting method and the number-fixed cycle-counting method) are suited to measuring the average frequency of a beat signal. Therefore, they are frequently used in nonlinear FMCW interference, such as sinusoidal-wave FMCW interference. Even in the linear FMCW interference, the average frequency is still useful because it can minimize the effects of nonlinear frequency response of the laser.

10.1.3 Pulse-Filling Method

For the very low-frequency beat signal, the cycle-counting methods are not suitable and the pulse-filling method is preferable. The pulse-filling method is based on the period measurement. The width of a gate signal is now referred to the period of the beat signal, and another high-frequency pulse signal is used to measure the gate width. For instance, if the first zero point of a beat signal in each modulation period is used to open a gate, and if the third zero point is used to close the gate, as shown in Figure 10.4, the period of the beat signal T_b will be

$$T_b = T_g$$
$$= MT_p , \tag{10-6}$$

where T_g is the gate width, M is the number of pulses filled in the gate, and T_p is the pulse period. The frequency of the beat signal v_b can be determined by

$$v_b = \frac{1}{MT_p} . \tag{10-7}$$

The maximum relative error $\Delta v_b / v_b$ of the pulse-filling method equals

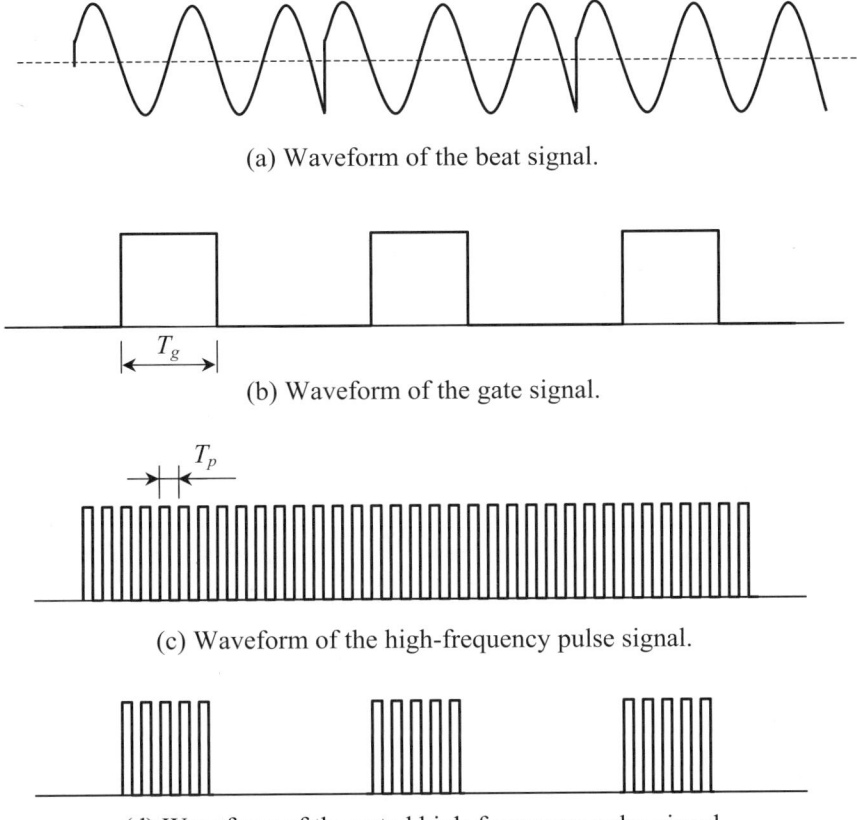

(a) Waveform of the beat signal.

(b) Waveform of the gate signal.

(c) Waveform of the high-frequency pulse signal.

(d) Waveform of the gated high-frequency pulse signal.

Figure 10.4. Pulse-filling method for frequency measurement.

$$\frac{\Delta \nu_b}{\nu_b} = \frac{\Delta M}{M} + \frac{\Delta T_p}{T_p}$$

$$= \frac{1}{M} + \frac{\Delta \nu_p}{\nu_p}$$

$$= \frac{\nu_b}{\nu_p} + \frac{\Delta \nu_p}{\nu_p} \ , \tag{10-8}$$

where ν_p is the frequency of the pulse signal and $\Delta \nu_p$ is the pulse frequency error. The first term in the equation is the quantification error, and the second term is the error caused by the inaccuracy of the pulse period.

The advantage of the pulse-filling method is that it uses only one pulse counter. The limitation is that the relative error will increase as the beat frequency rises. The minimum measurable frequency of the pulse-filling method equals the modulation frequency of the laser.

10.2 Phase Measurement

Phase measurement can also be performed with three methods: the pulse-filling method, the phase-locked-loop method, and the digital signal processing method. The first two methods can measure the phase difference between two sinusoidal signals of the same frequency. Obviously, they are suited to measuring the phase shift of the beat signal from a sawtooth-wave optical FMCW interferometer under the condition of a small measurement range. Usually, the beat signal from the sawtooth-wave FMCW interferometer is first sent to a dividing circuit to remove the intensity modulation, and then sent to an electric band-pass filter where the most intensive harmonic component of the beat signal is selected and its phase is compared with a standard reference signal, which normally is derived from the modulation signal for the laser. The digital signal processing method suits any kind of beat signal, but it is relatively complicated.

10.2.1 Pulse-Filling Method

Figure 10.5 illustrates the operating principle of the pulse-filling method. A standard sinusoidal reference signal (waveform (a)) and a phase-shifting sinusoidal beat signal (waveform (b)) are first converted into rectangular waves (waveforms (c) and (d)) by using Schmitt triggers. Then, these rectangular waves are used to trigger an R-S flip-flop. The output of the flip-flop is a rectangular wave (waveform (e)) whose gate width is determined by the phase difference between the input signal and the standard signal and is used to frame a high-frequency pulse signal (waveform (f)). Counting the number of pulses inside the gate (waveform (g)), the value of the phase difference can be determined.

Supposing the frequency of the pulse signal is N times as high as the frequency of the standard signal, the phase difference $\delta\phi$ between the input signal and the standard signal will be

$$\delta\phi = \frac{2\pi M}{N},$$

(10-9)

where M is the number of pulses filled in the gate and $\delta\phi$ is in radian. The maxi-

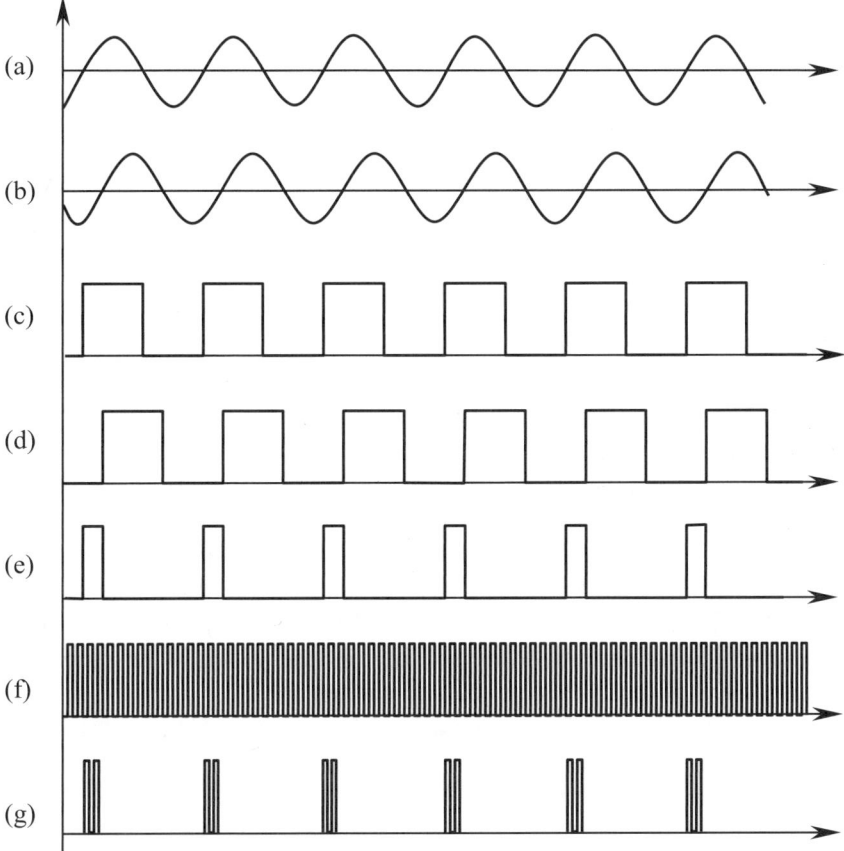

Figure 10.5. Pulse filling method for phase measurement.

mum phase error $\Delta\delta\phi$ of this method is determined by the fractional portion of the pulses and can be written as

$$\Delta\delta\phi = \frac{2\pi}{N}.$$

(10-10)

The pulse-filling method can measure a phase difference equal to or less than a period. For a bigger phase difference, the number of full periods can be determined by counting the pulses difference between the two comparison signals.

10.2.2 Phase-Locked-Loop Method

This method is based on the phase-locked-loop (PLL) technique. A typical phase-locked loop consists of four components: a phase-sensitive detector, a high-gain amplifier with a low-pass filter, a voltage-controlled oscillator (VCO), and an N-time frequency divider. The phase-sensitive detector outputs a dc voltage proportional to the phase difference between the two inputs, while the VCO varies the frequency of its sinusoidal wave oscillator according to the dc voltage. These components are usually connected in a loop arrangement, as shown in Figure 10.6, so that any phase difference between the input signal and the signal from the VCO is minimized.

Assume the frequency of the input signal is v_s and the frequency of the output signal from the VCO is Nv_0, where N is the coefficient of the frequency amplification of the VCO. If the phase of the input signal has a shift $\delta\phi$, the output signal from the phase-sensitive detector and amplifier makes the VCO increase its frequency to maintain the same phase as the input signal. If the phase shift is $2\pi/N$ radian, the output of the VCO will change by a cycle. Hence, measuring the change of the cycles, we can find the phase shift of the input signal.

Similarly, the maximum error $\Delta\delta\phi$ of this method is determined by the fractional portion of the cycles and can be written as

$$\Delta\delta\phi = \frac{2\pi}{N}.$$

(10-11)

The advantages of the phase-locked-loop method include high resolution and simple structure. (A single integrated circuit chip of the PLL is available.) The limitation is that the phase of the input signal cannot shift too fast.

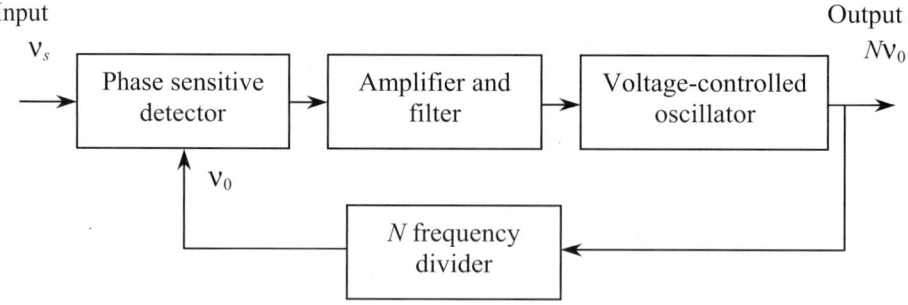

Figure 10.6. Phase-locked-loop method for phase measurement.

10.2.3 Digital Signal Processing Method

With the fast development of electronics and computer technology, high-performance analog-to-digital converters (A/D), high-performance digital-to-analog converters (D/A), high-speed specific digital processors, and high-speed microcomputers are already available. Therefore, the digital signal processing (DSP) method has become more practical and very important.

The digital signal processing method is the best way to measure the phase (as well as frequency) of a beat signal. Many defects and limitations of the analog methods discussed previously, such as in modulation waveform, phase-shift speed and measurement range, can be eliminated by using the digital signal processing approach.

In the digital signal processing method, the input signal is first digitized by a fast high-resolution A/D converter, and the amplitude and the phase of the signal are determined by high-speed computations in a microcomputer or a specific digital processor. The digital output may be used to display or be converted into an analog signal by a D/A converter sending it to a control system, as shown in Figure 10.7.

Digital signal processing, particularly real-time digital signal processing, is still a challenging topic. A lot of effort is still being put into this subject at present.

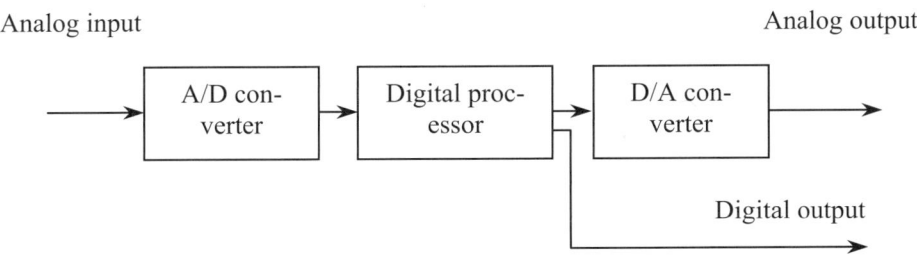

Figure 10.7. Principle of digital signal processing.

References

1. Hymans, A. J., and Lait, J., Analysis of a frequency-modulated continuous-wave ranging system, Proc. IEEE 107-B, 365–372 (1960).
2. Skolnik, M. I., *Introduction to radar systems*, McGraw-Hill, New York (1962).
3. Vali, V., and Shorthill, R. W., Fiber ring interferometer, Appl. Opt. 15, 1099–1100 (1976).
4. Cahill, R. F., and Udd, E., Phase-nulling fiber-optic laser gyro, Opt. Lett. 4, 93–95 (1979).
5. Goss, W. C., Goldstein, R., Nelson, M. D., Fearnehaugh, H. T., and Ramer, O. G., Fiber-optic rotation sensor technology, Appl. Opt. 19, 852–858 (1980).
6. Bergh, R. A., Lefevre, H. C., and Shaw, H. J., Single-mode fiber-optic polarizer, Opt. Lett. 5, 479–481 (1980).
7. Lefevre, H. C., Single-mode fibre fractional wave devices and polarization controllers, Electron. Lett. 16, 778–780 (1980).
8. Ulrich, R., Fiber-optic rotation sensing with low drift, Opt. Lett. 5, 173–175 (1980).
9. Hotate, K., Yoshida, Y., Higashiguchi, M., and Niwa, N., Fiber-optic laser gyro with easily introduced phase-difference bias, Appl. Opt. 20, 4313–4318 (1981).
10. Adams, A. J., *An introduction to optical waveguides*, John Wiley & Sons, New York (1981).
11. Bergh, R. A., Lefevre, H. C., and Shaw, H. J., All-single-mode fiber-optic gyroscope, Opt. Lett. 6, 198–200 (1981).
12. Bergh, R. A., Lefevre, H. C., and Shaw, H. J., All-single-mode fiber-optic gyroscope with long-term stability, Opt. Lett. 6, 502–504 (1981).
13. Arditty, H. J., and Lefevre, H. C., Sagnac effect in fiber gyroscopes, Opt. Lett. 6, 401–403 (1981).
14. Kobayashi, S., Yamamoto, Y., Ito, M., and Kimura, T., Direct frequency modulation in AlGaAs semiconductor lasers, IEEE J. of Quantum Electron. QE-18, 582–595 (1982).
15. Culshaw, B., and Giles, I. P., Frequency modulated heterodyne optical fiber Sagnac interferometer, IEEE J. of Quantum Electron. QE-18, 690–693 (1982).
16. Jackson, D. A., Kersey, A. D., Corke, M., and Jones, J. D. C., Pseudoheterodyne detection scheme for optical interferometers, Electron. Lett. 18, 1081–1083 (1982).
17. Giles, I. P., Uttam, D., Culshaw, B., and Davies, D. E. N., Coherent optical-fibre sensors with modulated laser sources, Electron. Lett. 19, 14–15 (1983).
18. Corke, M., Kersey, A. D., Jackson, D. A., and Jones, J. D. C., All-fibre Michelson thermometer, Electron. Lett. 19, 471–472 (1983).

19. Bergh, R. A., Lefevre, H. C., and Shaw, H. J., An overview of fiber-optic gyroscopes, J. of Lightwave Technol. LT-2, 91-107 (1984).
20. Uttam, D., Culshaw, B., Ward, J. D., and Carter, D., Interferometric optical fiber strain measurement, J. Physics E: Sci. Instrum. 18, 290–293 (1985).
21. Uttam, D., and Culshaw, B., Precision time domain reflectometry in optical fiber systems using a frequency modulate continuous wave ranging technique, J. of Lightwave Technol. LT-3, 971–976 (1985).
22. Uttam, D., Measurement of intermodal delay in a dual-mode optical fibre, Electron. Lett. 21, 1031–1033 (1985).
23. Franks, R. B., Torruellas, W., and Youngquist, R. C., Birefringent stress location sensor, Proc. SPIE 586, 84–89 (1985).
24. Thyagarajan, K., Bourbin, Y., Enard, A., Vatoux, S., and Papuchon, M., Experimental demonstration of TM mode-attenuation resonance in planar metal-clad optical waveguides, Opt. Lett. 10, 288–290 (1985).
25. Brooks, J. L. Wentworth, R. H., Youngquist, R. C., Tur, M., Kim, B. Y., and Shaw, H. J., Coherence multiplexing of fiber optic interferometric sensors, J. of Lightwave Technol. T-3, 1062–1072 (1985).
26. Franks, R. B., Torruellas, W., and Youngquist, R. C., An extended fiber optic stress location sensor, Opt. Acta 33, 1505–1518 (1986).
27. Economou, G., Youngquist, R. C., and Davies, D. E. N., Limitations and noise in interferometric systems using frequency ramped single-mode diode lasers, J. of Lightwave Technol. LT-4, 1601–1608 (1986).
28. Beheim, G., and Fritsch, K., Range finding using frequency-modulated laser diode, Appl. Opt. 25, 1439–1442 (1986).
29. Beheim, G., Fiber-optic interferometer using frequency-modulated laser diodes, Appl. Opt. 25, 3469–3472 (1986).
30. Mallallieu, K. L., Youngquist, R., and Davies, D. E. N., FMCW of optical source envelope modulation for passive multiplexing of frequency-based fibre-optic sensors, Electron. Lett. 22, 809–810 (1986).
31. Lefevre, H. C., Vatoux, S., Papuchon, M., and Puech, C., Integrated optics: a practical solution for the fiber-optic gyroscope, Proc. SPIE 719, 101–112 (1986).
32. Kersey, A. D., Dandridge, A., Phase-Noise reduction in coherence-multiplexed interferometric fibre sensors, Electron. Lett. 22, 616–617 (1986).
33. Sakai, I., Frequency-division multiplexing of optical-fibre sensors using a frequency-modulated source, Opt. Quantum Electron. 18, 279–289 (1986).
34. Sakai, I., Parry, G., and Youngquist, R. C., Multiplexing fiber-optic sensors by frequency modulation: cross-term considerations, Opt. Lett. 11, 183–185 (1986).
35. Sakai, I., Youngquist, R. C., and Parry, G., Multiplexing of optical fiber sensors using a frequency-modulated source and gated output, J. of Lightwave Technol. LT-5, 932–939 (1987).
36. Dyott, R. B., Bello, J., and Handerek, V. A., Indium-coated D-shaped-fiber polarizer, Opt. Lett. 12, 287–289 (1987).
37. Brooks, J. L., Moslehi, B., Kim, B. Y., and Shaw, H. J., Time-domain addressing of remote fiber-optic interferometric sensor arrays, J. of Lightwave Technol. 5, 1014–1023 (1987).
38. Den Boef, A. J., Interferometric laser rangefinder using a frequency modulated diode laser, Appl. Opt. 26, 4545–4550 (1987).

39. Leilabady, P. A., Optical fiber point temperature sensor, Proc. SPIE 838, 231–237 (1987).
40. Kubota, T., Nara, M., and Yoshino, T., Interferometer for measuring displacement and distance, Opt. Lett. 12, 310–312 (1987).
41. Chen, J., Ishii, Y., and Murata, K., Heterodyne interferometry with a frequency-modulated laser diode, Appl. Opt. 27, 124–128 (1988).
42. Farahi, F., Gerges, A. S., Jones, J. D. C., and Jackson, D. A., Time-division multiplexing of fibre optic interferometric sensors using a frequency modulated laser diode, Electron. Lett. 24, 54–55 (1988).
43. Farahi, F., Jones, J. D. C., and Jackson, D. A., Multiplexed fibre-optic interferometric sensing system: combined frequency and time division, Electron. Lett. 24, 409–410 (1988).
44. Francis, D. M., Effect of laser coherence and system design on FMCW multiplexed sensor system performance, Proc. SPIE 1169, 159–171 (1989).
45. Meggitt, B. T., and Palmer, A. W., A fibre optic compatible signal-processing scheme for dual wavelength interferometry using Fourier harmonics, Measurement 7, 50–54 (1989).
46. Sorin, W. V., Donald, D. K., Newton, S. A., and Nazarathy, M., Coherent FMCW reflectometry using a temperature tuned Nd:YAG ring laser, IEEE Photonics Technol. Lett. 2, 902–904 (1990).
47. Berkoff, T. A., and Kersey, A. D., Interferometric fibre displacement/strain sensor based on source coherence synthesis, Electron. Lett. 26, 452–453 (1990).
48. Venkatesh, S., and Dolfi, D. W., Incoherent frequency modulated CW optical reflectometry with centimeter resolution, Appl. Op. 29, 1323–1326 (1990).
49. Kotrotsios, G., Benech, P., and Parriaux, O., Multipoint operation of two-mode FMCW distributed fiber-optic sensor, J. of Lightwave Technol. 8, 1073–1077 (1990).
50. Lu, Z. J., and Blaha, F. A., A two-mode fiber optic strain sensor system for smart structures and skins, Proc. SPIE 1370, 180–188 (1990).
51. Hogg, D., Janzen, D., Valis, T., and Measures, R. M., Development of a fiber Fabry-Perot strain gauge, Proc. SPIE 1588, 300–307 (1991).
52. Venkatesh, S., and Sorin, W. V., Fibre-tip displacement sensor using sinusoidal FM-based technique, Electron. Lett. 27, 1652–1654 (1991).
53. Zheng, G., Tian, Q., and Liang, J. W., Multifunction multi-channel remote-reading optical fiber sensor system, Proc. SPIE 1572, 299–303 (1991).
54. Zheng, G., Tian, Q., and Liang, J. W., Frequency division multiplexing optical fiber displacement sensor with high precision, Acta IMEKO 1991, 1413–1418 (1991).
55. Toyama, K., Fesler, K. A., Kim, B. Y., and Shaw, H. J., Digital integrating fiber-optic gyroscope with electronic phase tracking, Opt. Lett. 16, 1207-1209 (1991).
56. Chien P. Y., and Pan, C. L., Multiplexed fiber-optic sensors using a dual-slope frequency-modulated source, Opt. Lett. 16, 872–874 (1991).
57. Ishii, Y., and Onodera, R., Two-wavelength laser-diode interferometry that uses phase-shifting techniques, Opt. Lett. 16, 1523–1525 (1991).
58. Keiser, Gerd, *Optical Fiber Communication*, McGraw-Hill, New York (1991).
59. Berkoff, T. A., and Kersey, A. D., Reflectometric two-mode elliptical-core fiber strain sensor with remote interrogation, Electron. Lett. 28, 562–564 (1992).

60. Amann, M. C., Phase noise limited resolution of coherent lidar using widely tunable laser diodes, Electron. Lett. 28, 1694–1696 (1992).

61. Passy, R., Gisin, N., and von der Weid, J. P., Mode hopping noise in coherent FMCW reflectometry, Electron. Lett. 28, 2186–2188, (1992).

62. Venkatesh, S., Sorin, W. V., Phase noise considerations in coherent optical FMCW reflectometry, J. of Lightwave Technol. 11, 1694–1700 (1993).

63. Dieckmann, A., FMCW-lidar with tunable twin-guide laser diode, Electron. Lett. 30, 308–309 (1994).

64. Dieckmann, A., and Amann, M. C., Phase-noise-limited accuracy of distance measurements in a frequency-modulated continuous-wave lidar with a tunable twin guide laser diode, Opt. Eng. 34, 896–903 (1995).

65. Onodera, R., and Ishii, Y., Two-wavelength laser-diode interferometer with fractional fringe techniques, Appl. Opt. 34, 4740–4746 (1995).

66. Ishii, Y., and Onodera, R., Phase-extraction algorithm in laser-diode phase-shifting interferometry, Opt. Lett. 20, 1883–1885 (1995).

67. Onodera, R., and Ishii, Y., Two-wavelength laser-diode heterodyne interferometry with one phasemeter, Opt. Lett. 20, 2502–2504 (1995).

68. Bass, M., *Handbook of Optics, Volume I, 2nd edition*, McGraw-Hill, New York (1995).

69. Bass, M., *Handbook of Optics, Volume II, 2nd edition*, McGraw-Hill, New York (1995).

70. Christiansen, D., *Electronics Engineers' Handbook, 4th edition*, McGraw-Hill, New York (1996).

71. Zheng, G., Campbell, M., and Wallace, P. A., Length-division-sensitive, birefringent fiber FMCW remote strain sensor, Proc. SPIE 2783, 307–311 (1996).

72. Zheng, G., Campbell, M., Wallace, P. A., and Holmes-Smith, A. S., Single-piece-fiber FMCW remote strain sensor with environment-insensitive lead-in lead-out fibers, Proc. SPIE 2839, 272–276 (1996).

73. Zheng, G., Campbell, M., and Wallace, P. A., Reflectometric frequency modulation continuous wave distributed fiber optic stress sensor with forward coupled beams, Appl. Opt. 35, 5722–5726 (1996).

74. Zheng, G., Campbell, M., Wallace, P. A., and Holmes-Smith, A. S., A practical birefringent fiber Sagnac ring force sensor, Proc. SPIE 2895, 196–200 (1996).

75. Zheng, G., Campbell,M., Wallace,P. A., and Holmes-Smith, A. S., FMCW birefringent fiber strain sensors based on Sagnac rings, Proc. SPIE 2837, 177–182 (1996).

76. Zheng, G., Campbell, M., Wallace, P. A., and Holmes-Smith, A. S., Distributed FMCW reflectometric birefringent fiber stress sensor, Proc. SPIE 2838, 291–295 (1996).

77. Zheng, G., Campbell, M., Wallace, P. A., and Holmes-Smith, A. S., Configurations of remote birefringent fiber strain sensors using a frequency modulation continuous wave technique, Appl. Opt. and Opto-Electron. 1996, 380–385 (1996).

78. Zheng, G., Campbell, M., and Wallace, P. A., Sagnac birefringent fiber strain sensor with FMCW technique, Proc. SPIE 2784, 102–105 (1996).

79. Campbell, M., Zheng, G., Wallace, P. A., and Holmes-Smith, A. S., Distributed stress sensor with a birefringent fiber Sagnac ring, Proc. SPIE 2838, 138–142 (1996).

80. Campbell, M., Zheng, G., and Wallace, P. A., Birefringent fiber remote strain sensor with FMCW interferometry, Proc. SPIE 2784, 98–101 (1996).

81. Campbell, M., Zheng, G., Wallace, P. A., and Holmes-Smith, A. S., Reflectometric birefringent fiber sensor for absolute and relative strain measurement, Proc. SPIE 2839, 254–259 (1996).

82. Campbell, M., Zheng, G., Wallace, P. A., and Holmes-Smith, A. S., Reflectometric birefringent fiber absolute and relative strain sensor with environment-insensitive lead-in/lead-out fiber, Proc. SPIE 2895, 222–227 (1996).

83. Campbell, M., Zheng, G., and Wallace, P. A., FMCW birefringent fiber strain sensor with two forward-coupled beams, Proc. SPIE 2783, 312–315 (1996).

84. Liyama, K., Wang, L. T., and Hayashi, K., Linearizing optical frequency-sweep of a laser diode for FMCW reflectometry, J. of Lightwave Technol. 14, 173–178 (1996).

85. Minoni, U., Scotti, G., and Docchio, F., Wide-range distance meter based on frequency modulation of an Nd:YAG laser, Opt. Eng. 35, 1949–1952 (1996).

86. Zhou, X. Q., Liyama, K., and Hayashi, K., Extended-range FMCW reflectometry using an optical loop with a frequency shifter, IEEE Photonics Technol. Lett. 8, 248–250 (1996).

87. Campbell, M., Zheng, G., Wallace, P. A., and Holmes-Smith, A. S., A distributed FMCW fiber stress sensor based on a birefringent Sagnac ring configuration, Opt. Rev. 4, 114–116 (1997).

88. Takahashi, Y., Yoshino, T., Ohde, N., Amplitude-stabilized frequency-modulated laser diode and its interferometric sensing applications, Appl. Opt. 36, 5881–5887 (1997).

89. Nérin, P., Labeye, P., Besesty, P., Puget, P., Chartier, G., Bergeon, M., FMCW technique using self-mixing inside a LiTaO$_3$-Nd:YAG microchip laser for absolute distance and velocity measurements, Proc. SPIE 3100, 144–151 (1997).

90. Rovati, L., Minoni, U., Docchio, F., Dispersive white light combined with a frequency-modulated continuous-wave interferometer for high-resolution absolute measurements of distance, Opt. Lett. 22, 850–852 (1997).

91. Campbell, M., and Zheng, G., A novel fibre optic strain sensor, International J. of Electron. 85, 545–552 (1998).

92. Ghatak, A. K., and Thyagarajan, K., *Introduction to fiber optics*, Cambridge University Press, Cambridge (1998).

93. Campbell, M., Zheng, G., Holmes-Smith, A. S., and Wallace, P. A., A frequency-modulated continuous wave birefringent fibre-optic strain sensor based on a Sagnac ring configuration, Meas. Sci. Technol. 10, 218–224 (1999).

94. Karlsson, C. J., and Olsson, F. Å. A., Linearization of the frequency sweep of a frequency-modulated continuous-wave semiconductor laser radar and the resulting ranging performance, Appl. Opt. 38, 3376–3385 (1999).

95. Dalton, S. D., Fourier coefficients for range identification in FMCW radar systems, Proc. SPIE 3704, 28–35 (1999).

96. Minoni, U., and Rovati, L., High-performance front-end electronics for frequency-modulated continuous-wave interferometers, IEEE Trans. Instrum. Meas. 48, 1191–1196 (1999).

97. Ishii, Y., and Takahashi, T., Laser-diode phase-conjugate interferometry with a frequency-modulated continuous-wave technique, Proc. SPIE 4110, 55–63 (2000).

98. Seah, L. K., and Won, P. C., Distributed FMCW reflectometric optical fiber strain sensor, Proc. SPIE 4416, 66–69 (2001).

99. Schneider, R., Thürmel, P., Stockmann, M., Distance measurement of moving objects by frequency modulated laser radar, Opt. Eng. 40, 33–37 (2001).

100. Dupuy, D., Lescure, M., and Tap-Beteille, H., FMCW laser range-finder with an avalanche photodiode working as an optoelectronics mixer, Proc. SPIE 4546, 54-63 (2002).

101. Dupuy, D., and Lescure, M., Improvement of the FMCW laser range-finder by an APD working as an optoelectronics mixer, IEEE Trans. on Ins. and Meas. 51, 1010-1014 (2002).

102. Amann, M. C., Bosch, T., Lescure, M., Myllylä, R., and Rioux, M., Laser ranging: a critical review of usual techniques for distance measurement, Opt. Eng. 40, 10–19 (2002).

103. Won, P. C., Seah, L. K., and Xie, G. P., Quasi-distributed frequency-modulated continuous-wave reflectometric optical fiber strain sensor, Opt. Eng. 41, 788–795 (2002).

104. Hecht, E., *Optics, 4th edition*, Addison-Wesley, Reading, MA (2002).

105. Zheng, J., Single-mode fibre frequency-modulated continuous-wave Sagnac gyroscope, Electron. Lett. 40, 1255–1257 (2004).

106. Zheng, J., Differential birefringent fiber frequency-modulated continuous-wave Sagnac interferometer, in *CLEO/IQEC 2004*, San Francisco, California, Sponsored by APS, IEEE-LEOS and OSA, paper CThII3 (2004).

107. Zheng, J., Analysis of optical frequency-modulated continuous-wave interference, Appl. Opt. 43, 4189–4198 (2004).

Index

Springer Series in
OPTICAL SCIENCES

Springer Series in
OPTICAL SCIENCES